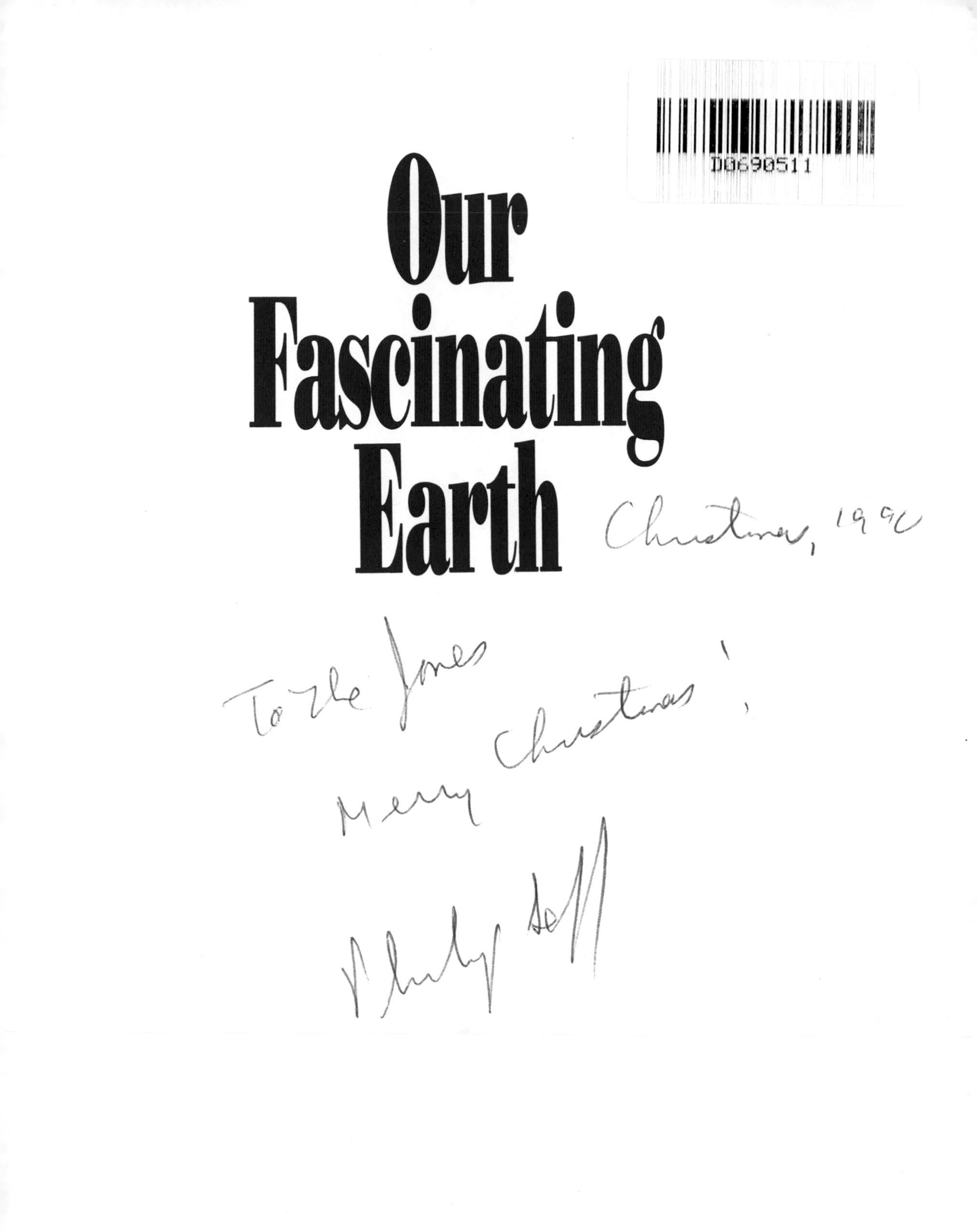

Our Fascinating Earth

Our Fascinating Earth

Strange, True Stories of Nature's Oddities, Bizarre Phenomena, and Scientific Curiosities

Philip Seff, Ph. D., and Nancy R. Seff, M.Ed.

CB
CONTEMPORARY
BOOKS
CHICAGO

Library of Congress Cataloging-in-Publication Data

Seff, Philip.
Our fascinating earth : strange true stories of nature's oddities, bizarre phenomena, and scientific curiosities / Philip Seff and Nancy R. Seff.
p. cm.
Includes bibliographical references.
ISBN 0-8092-4185-4 : $12.95
1. Natural history—Miscellanea. 2. Curiosities and wonders.
I. Seff, Nancy R. II. Title.
QH45.5.S428 1990
508—dc20 90-30628
CIP

To our parents:
who showed us how to live

Interior illustrations by Mel Chadwick and Doug Wellons

Published by Contemporary Books, Inc.
180 North Michigan Avenue, Chicago, Illinois 60601
Manufactured in the United States of America
International Standard Book Number: 0-8092-4185-4 (paper)
0-8092-4073-4 (book club)

CONTENTS

CHAPTER THREE

CHAPTER FOUR

CHAPTER FIVE

CHAPTER SIX

CHAPTER SEVEN

CHAPTER EIGHT

CHAPTER NINE

INTRODUCTION

During my experiences as a professor of geology I was routinely perplexed by the reluctance of many students to take a laboratory science, despite the fact that a year of science was required by each institution. Many put off satisfying their science requirement until their final year. I agonized over how to present science to the ordinary non-science student so that, far from being the collegiate bogeyman, it would become interesting and entertaining as well as educational and practical.

The method came to me during the middle 1970s when a colleague suggested that I write up some of the interesting facts of science for publication as a newspaper feature. Being a professional geologist, I was naturally inclined in the direction of earth science and called the newly created illustrated panels "Geology Sketchbook."

An agent in New York reviewed the panels and promptly notified me that the subject matter was too restricted for the feature ever to become a crowd pleaser. I recognized the idea as sound, because not everyone can be compelled to radiate enthusiasm for geology. The name was changed to "Our Fascinating Earth," and it was redesigned to include all phases of science. This involved an extraordinary amount of research, all of which was most enjoyable. It was like returning to school for only the interesting subjects.

Realizing that almost everyone is interested in some phase of science, I planned the daily panels so they would be varied. By dealing with a different facet of science each day, I was bound eventually to collide with a particular interest of every reader. Then it would become only a matter of time before readers, perusing the panel to find out if it contained something of interest, would discover that their realm of interests had become broadened.

This being an era of science and trivia, I was, without knowing it, combining these two ingredients into scientific trivia. The variation of subject matter became a permanent format and was very effective. Readers have often told me that when they come home from work or school and sit down to relax, the first thing they do is pick up their copy of the evening paper and turn to the page containing the panel of "Our Fascinating Earth" as they wonder, "What did Phil write about today?" They have told me that it gave them a morsel of unusual science information while entertaining them.

With so many natural wonders at my disposal I have never had to invade the realm of science fiction. Every panel ever published was carefully researched as fact. If there were doubts about the validity of the data, I would state them if the subject were sufficiently important or discard questionable material. Being myself simply "fascinated" by the unusual in science, I am pleased that truth is stranger than fiction and that I have never been compelled to compromise the facts. I have sought to make people more aware of the world around them and to enhance their appreciation and understanding of natural wonders. I trust that, in some humble way, I have succeeded.

It didn't take long, from the inception of the feature with Copley News Service, for "Our Fascinating Earth" to spread into newspapers nationwide. My original goal of proving that science is of interest to everyone was quickly realized. My correspondents inform me that teachers at all levels use the panels as teaching aids, especially in elementary grades. It is not unusual for teachers to post them on bulletin boards and use them as a reference or to give assignments based on the phase of science discussed in that particular panel.

Eventually I began to get letters from educators requesting or ordering a book of "Our Fascinating Earth." This was impossible, for until 1980 I hadn't even considered putting the material into book form. Other requests, just as numerous, were for "more information on the subject." Being constantly reminded of the mountain of data that I had amassed, which was going nowhere, I decided to write the book.

The prescription for the book's format was that it follow the same direction as the daily newspaper panel. Material for the book was drawn from the panels, but since they gave only a limited amount of information the subject of each panel was researched and written more fully. By varying the subject matter for the book as I have in the newspaper feature, I have tried to provide something of interest for everyone. As I have discovered from dilettantish readers of the manuscript, the subject matter changes just when the readers were ready for something else. Typical comments were that the highly varied nature of each chapter kept the readers' interest recharged to the extent that they had difficulty putting the book down.

Most books of this nature, which provide morsels of information on many subjects, are arranged by subject matter. This is both logical and scientific. But the recurrent pattern of "Our Fascinating Earth" has been a potpourri of science. Hence, to avoid a consistent maze of subject matter, the book has been broken down into chapters of nearly uniform length. When a reader has arrived at the end of a chapter, his or her mind has been able to focus on a dozen or more different topics. At this point the chapter's end becomes a natural place for the reader to pause and reflect.

My wife, Nancy, a librarian and teacher, has assisted in many positive ways in the making of this book. Not only has she edited and refined everything that was written, but her research skills are evident everywhere. Being intensely interested in the earth and its wonders, she has worked independently on several articles included in this book. With evidence of her handiwork on every page, it is only natural that her name appear on the title page as coauthor.

We are most grateful to the numerous people who have checked the accuracy of our data and offered many helpful suggestions. We recognize Mr. Rick Right, who continuously read the developing manuscript and gave invaluable advice. We especially wish to thank Dr. Judson Sanderson and Dr. Harvey Wieseltier, who gave freely of their time and expertise to edit this book.

We further appreciate the exacting criticism of the final manuscript by Ms. Harriet Garfinkle and Dr. Ron Seff, whose work has reduced the margin for error or misstatement to the lowest level.

At the risk of sounding corny, I offer the reader a chance to enter into the highly varied, fascinating realm of nature, surely an experience worth having. So enjoy, enjoy!

Philip Seff
Redlands, California

CHAPTER ONE

Krakatoa

Until nearly the end of the nineteenth century volcanology had hardly achieved the status of a science. But then came the great explosion of . . . Krakatoa.

Part of a group of volcanic islands, Krakatoa lay in the Sunda Strait between Java and Sumatra, a zone where three fissures of the earth's crust intersect.

Krakatoa was long recognized as the remnants of an old volcano, but its three craters had not been active for over 200 years and were believed to be definitely extinct. It was on May 20, 1883, that explosive sounds were first heard coming from one of Krakatoa's craters. These noises aroused some curiosity, but nobody was alarmed. In fact a few days later an excursion steamer brought a party of sightseers to the island. Some of the tourists climbed the largest peak and stood there in awe as they observed a steady column of smoke roaring furiously out of a vent more than thirty yards wide.

This violent activity increased rapidly during the next few months. Great quantities of smoke and ash spewed from the 2,600-foot volcano, often obscuring the midday sunlight. This was the prelude to one of the most gigantic explosions recorded in modern history.

An earthquake had opened a fissure, allowing seawater to flow into the magmatic chambers of the volcano. Mingling with the intense heat of the molten rock, the water quickly became steam. This set off eruptions that formed dense clouds of fine volcanic dust and steam visible for hundreds of miles at sea.

On August 27, 1883, several explosions occurred, each more devastating than the last. The third, estimated as the equivalent of 20,000 megatons, utterly destroyed the island, leaving a hole to a depth of 1,000 feet below sea level where three mountains had stood.

The explosions generated huge tidal waves (tsunami), which traveled

across the sea at over 400 miles per hour. When these deadly waves struck the land, they swept away more than 1,000 villages on neighboring islands and killed well over 36,000 people. Thirty-two hours later and halfway around the world, the much weakened wave reached the English Channel.

Dust was shot into the upper atmosphere and blown around the world, causing spectacular dawns and sunsets for over three years after the explosion. Pumice was cast into the sea in such enormous quantities that shipping was slowed considerably. Many bays and harbors became choked with thick floating "rafts" of light, frothy lava.

Krakatoa's Big Bang ranged over one-thirteenth of the earth's surface and was heard by hundreds of thousands, most of whom thought it was a ship at sea firing distress cannons. At Timor Island, 1,351 miles away, government boats were put out to sea to investigate the cannonading. In New Guinea, 1,800 miles away, natives asked the resident missionary why the white people were firing cannons at sea.

Just how high the tsunami were has been a subject of much controversy. The only factor agreed on by every source is that they were gigantic. The height of the waves quite naturally varied with the configuration of the shorelines on which they struck.

One definite record of the wave exists, for a section of a Java beach. A village once stood on this beautiful seacoast, where the sea was confined in a large, sheltered cove flanked by tall mountains. A 200-foot cliff overlooked the village, and the inhabitants built a long, winding road up the cliffside to the top, where they enjoyed a spectacular view of the sea.

On that fateful August 27 most of the villagers had climbed to the top of the cliff to observe the great eruption. When the sea suddenly withdrew several thousand feet from its former shoreline, the remaining villagers, anticipating what was going to happen, fled up the cliff. Most of them had reached the top when the sea returned in a sudden landward rush. A gigantic wall of water swept over the village and crashed against the cliff face. A group of six people had been observed near the top when the wall of water smashed over them and roared up the cliff at least another forty feet. After the waters receded, all six were gone. The survivors at the top noted that when the tidal wave struck, the six people had been standing on the road at least 140 feet above the base of the cliff!

As great as the explosion of Krakatoa was, scientists estimate that it was only one-fifth the size of the eruption of Thera (Santoríni) in the Aegean Sea in 1470 B.C. So great was this eruption that it permanently destroyed the centers of the Minoan civilization!

• •

He Called in Trouble

It is not unusual for human hunters to use imitative sounds to lure prey within range of their guns. Among large carnivores there are also a few that make use of imitative sounds for hunting; the tiger is one of them. It is very efficient in imitating the call of a stag and is able to entice deer to come within reach of its fangs and claws. This type of hunting can become complicated, however, particularly when no one has determined who is the hunter and who the hunted. An incident that occurred during the 1930s in the forests of India attests eloquently to this type of confusion.

A native hunter was using a stag horn to try to bring in a deer. His prey answered with, he noted, an unusually clear voice. The man slowly moved in the direction of the "deer," stopping on occasion to blow his horn and keep the prey interested. The prey continued to give a deerlike response and was getting noticeably closer.

And so the stalking continued, with the hunter stopping and calling periodically and always receiving an answer. Finally the hunter moved into a clearing. There he froze in midstep; stalking toward him and still in good voice was the animal he had called in—a full-grown tiger!

Fortunately the tiger was not a man-eater, although it is very likely that for a short time he considered becoming one. Voicing his displeasure in undisguised tiger tones, the tiger turned and walked to the edge of the clearing, where he stopped and again snarled his vexation. The man fainted. When he regained his senses, he was alone; the tiger had gone.

• •

Life in a Harsh Environment

Plants and animals that live in the desert have evolved excellent adaptations to their harsh, relentless environment. The saguaro and mesquite serve as unique but typical examples for the plant kingdom.

The largest of the known cactus species, the saguaro is found in the Sonoran Desert of southern Arizona and northern Mexico. It can reach heights of up to fifty feet in its over 200 years of growth. Its shallow roots extend radially, like interlaced wire spokes, as far as sixty-five feet in all directions around the plant. These roots form an effective plumbing system for the saguaro. When it rains, the network of roots sponges up water at a rapid rate, pumping it into the plant, where it is distributed to all parts of the giant cactus. The root system is so effective that after a good storm it can send enough water into the saguaro for the cactus to get along without another drink for as long as four years.

The saguaro's accordion-pleated stem permits it to expand and contract, a critical factor in its adaptation to the desert. After a summer cloudburst the saguaro may absorb as much as a ton of water; this is made possible by the expansion properties of the pleated stem. A mature saguaro may weigh over ten tons, of which four-fifths may be water. The cactus is then well prepared for the dry months that will follow.

The mesquite, a rather common tree in the deserts of the American

Southwest, is very well adapted to such arid environments. It has an extensive root system that often reaches almost sixty feet below the surface of the land. The roots are constantly searching for and tapping zones of underground water. These roots become so thick and tough that many can be used for firewood.

The mesquite's crowning achievement occurs just after birth, because it will actually delay growth above ground until its roots have located water! Then, and only then, will the plant begin its surface growth, for now it is adequately equipped to face the harsh desert environment.

For those animals that have adapted to an arid environment, the food they eat is their most important source of water. The water content of most desert plants is as high as 90 percent, making them a critical source of water for plant-eating animals. An herbivorous animal such as the desert wood rat fulfills almost all of its water needs by eating cactus. The wood rat itself is also made up mostly of water, so predators such as the gopher snake can obtain most of the moisture they need from a diet of rodents.

Perhaps the most successful adapter to the desert is the kangaroo rat of the American Southwest. It never drinks water! Subsisting on dry plants and seeds, it gets its fluids almost entirely from the water of oxidation, which is produced during the digestive process. An average kangaroo rat can produce fifty-four grams of water from about one hundred grams of plant seeds.

Of all desert dwellers, humans are the least able to survive in an arid environment. The human body has evolved no water-storing organs or similar aids to endure desert conditions. To survive comfortably on a hot day in the Sahara, those who are not accustomed to the desert should drink about six quarts of water. This is not always feasible, and every year a number of tourists die in the great Sahara Desert.

A few years ago in the Egyptian desert five tourists tried to reach an oasis 125 miles distant in three vehicles, all unfit for desert travel. To add insult to their rashness, they completely ignored the recommendation of six quarts of water per person and carried only one quart each.

They did manage to drive ninety-five miles before the cars, choked with dust and/or stuck in the sand, could travel no further. The men then set out on foot, retracing their tracks to reach the point of origin rather than following the trackless sands to their destination just thirty miles away.

A few days later a desert patrol found their abandoned automobiles and followed the tire tracks. Not one of the tourists had survived. The first body was discovered only ten miles from the abandoned vehicles. The farthest any man had gotten before collapsing was thirty miles. The toughest of the group was a little poodle whose mummified body was found fifty miles from the cars; it had obviously been headed back in the right direction.

The prescribed six quarts of water is based merely on an average, as individuals vary greatly in their water requirements. Those whose body hydration of water is extremely low should never venture out into the desert, because getting stranded could be fatal if they became dehydrated before becoming aware of it.

An incident that occurred in the Sahara in August 1973 is a tragic example. A couple traveling from one oasis to another developed engine trouble, and the man left the woman sitting in the car shade while he went for help. Securing assistance, he returned by car only five hours later. He found her still sitting where he had left her—but she had perished from the effects of unrelenting heat and thirst!

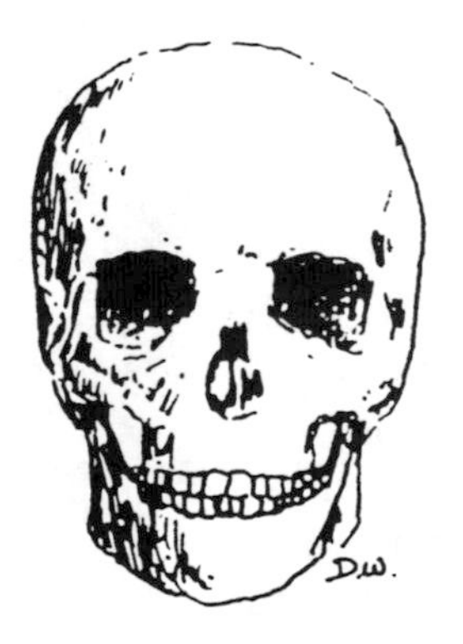

A Casualty of the Sun

The cyclic increase in solar flare activity has resulted in one major casualty to date—the satellite known as Skylab.

The sudden, intense rise in solar flares caused the earth's atmosphere to heat up, which made it puff out like a balloon into Skylab's orbital path. This in turn produced a braking effect on the American satellite. When Skylab slowed down, its orbit began to decay rapidly, and it fell to earth prematurely.

Skylab was the first American space station, launched by NASA on May 14, 1973. Its purpose was to provide an orbiting space laboratory in which astronauts could test human abilities to live and work in outer space. The space laboratory was expected to remain safely in orbit for at least ten years. With increased technical advances available by then, scientists had planned to move the satellite to a safer orbit. However, they had not calculated the effects of increased solar activity.

And so, on July 11, 1979, after orbiting the earth over 31,000 times, the American space laboratory plummeted to earth in thousands of blazing pieces. Fortunately none of the falling debris caused any injuries or property damage, as most of the fragments fell into the Indian Ocean and a desolate area of southwestern Australia.

• •

A Game of Bluff

Movies and television often depict the gorilla as a fearsome, ferocious beast, the villain of many fantasies. But that is all these portrayals are—fantasies. The wild antics of this frequently misrepresented animal—the hooting, screaming, flailing arms, and chest beating—are in reality unmitigated bluff.

A scientist conducting research on gorillas in the wild reported that on several encounters when he startled a gorilla the animal would put on a tremendous show. Each time there were unfinished charges, accompanied by great screaming and chest beating. The animal would interrupt its histrionics periodically and peer inquisitively at the scientist to see whether they had had any effect. Seeing no reaction, it would finally turn and walk away to resume its feeding.

The gorilla seems to carry the game of bluff to ridiculous extremes. Researchers during the 1980s observed a dispute between two large males who, with much roaring and chest beating, rushed furiously at each other. Then both stopped suddenly and stared silently into each other's eyes, their

faces only a foot apart. After about thirty seconds of stare-down, with neither giving ground, they both turned and ambled away without so much as a backward glance.

• •

Odd Marriage Customs

The courting habits of primitive people often involve complicated rituals. In Paraguay when two women of the Chaco Indian tribe want to marry the same man, they don tapir-skin boxing gloves and fight it out. The winner gets the man, who, incidentally, has very little to say in the matter.

The women of the Triobrian Islands near Papua have quite a different approach to marriage. They simply go up to the man of their choice and bite him on the arm; his opinions and preferences are again of little consequence.

The ancient Babylonians were quite practical when it came to marriage. They never waited for any of their women to become "old maids"; it was customary to auction off all marriageable females annually. Men bid very high for the most attractive females. The money paid for them was used strictly as dowries for the less attractive women for whom no one had bid. With such fringe benefits, an unattached woman was a notable rarity.

The ancient Greeks had a marital custom that, if applied today, would make old age obsolete. The women apparently managed to remain young by counting their age from the date on which they were married rather than from the day of their birth. They felt that the wedding marked the real beginning of a woman's life; all that went before was merely preparation. This usually gave a woman an extra twenty years of youth. The average life span of the ancient Greek woman was about fifty years; yet it was rare to find a woman over "thirty."

Years ago bizarre courting rituals were enforced strictly to keep the bride's virtue intact before the actual marriage. And in some parts of the world the methods were extreme. In the Solomon Islands a bride-to-be was kept in a cage, closely guarded, and not released until the time of the wedding. The girl's parents would keep an unusually sharp eye on their future son-in-law, who had to account for his whereabouts at all times.

In Wales future grooms had to develop artistic skill if they wished to be allowed to visit their brides-to-be. To keep the grooms' hands busy until the wedding, they had to make wooden spoons of very elaborate and delicate design for the girls' parents.

In contrast, nineteenth-century Scottish law required brides to certify their productivity by being pregnant on their wedding day. The law was enforced.

Standards of beauty have changed over the ages as well. Nowadays Americans consider a thin, shapely woman sexy and desirable. But it wasn't always so. In the late nineteenth century the great American beauty was Lillian Russell, and many a young man sighed over her photographs. This famous singer and actress, at the peak of her career, topped the scales at 186 pounds! The "fat is beautiful" viewpoint still prevails, not in America but in parts of Nigeria. Here, when young girls reach puberty, they enter fattening houses, where they spend their time eating almost constantly. When they emerge months later, they appear as "mountains of flesh" and are then considered fit for marriage.

Unusual methods of ending "marital bliss" have also been recorded. Back in the 1870s, in the city of Corinne, Utah, divorce was made so simple that any man could obtain one instantly. By merely slipping a $2.50 gold coin into a machine and turning a crank, he received divorce papers already signed by the local judge. But only men qualified for obtaining a divorce in this manner. The machine was extremely popular—for a while. Utah statutes failed to back up these slot machine divorces, and they were later declared illegal. As a result many men found they had unwittingly become bigamists.

• •

Robber Flies

Some insects have taken on a strange relationship with spiders: they steal the food stored by spiders in their webs.

One species of dance fly hovers near the edge of orb spider webs and lands carefully on tiny insects that have been entangled in the web but are too small for the spider to notice.

Another type of insect, the scorpion fly, also lands on spider webs to steal the entrapped prey. But the scorpion fly plays a dangerous game. It is too large to go unnoticed by the resident spider, so it employs an unusual weapon. The invader's tactic is to regurgitate a liquid that both prevents the fly from becoming entangled in the sticky glue on the web's thread and repels the attacking spider. This technique is, of course, risky, and if the spider attack succeeds, the robber fly becomes an additional victim. One noted entomologist estimates that the major cause of death for scorpion flies is predation by web-building spiders.

A species of flies of the family Chloropidae has evolved a more effective stratagem. It spends much time resting near an orb weaver that hides its web in a curled leaf. Any insect landing on the web can expect to be killed by the spider, wrapped in silk, and stored for a later meal. The fly stealthily closes in on the bundle of silk, avoiding entanglement in the web, and, when it is close enough, unfolds a long, sucking proboscis and gently sticks it into the prey. A minute or two later the fly flutters away with its abdomen bulging, leaving the spider with an empty shell of a meal.

Another species of robber fly actually sits right on the dorsal part of the spider's body, where the spider cannot brush it off. When the spider is

feeding, the fly rushes in, sucks up a quick meal, and hastily returns to its position on the spider's body.

An Australian species belonging to the same family has added a symbiotic twist to its relationship with spiders. It has apparently reached an understanding with the spider: in payment for a meal pilfered from its host, the fly cleans the spider after a meal. It carefully mops the spider's mouth and fangs and even throws in an extra service by cleaning waste matter from around its anal area.

Incidents at Shanidar Cave

Scientists have discovered that the human emotions of compassion, hate, and love are probably as old as man himself. The excavations of Shanidar Cave, Iraq, during the 1950s unearthed more than just eight skeletons of Neanderthal man. They clearly showed that by the middle of the Paleolithic Stone Age, 60,000 years ago, human beings were capable of strong emotions unexpected of the savage prehistoric caveman.

The first skeleton exhumed in 1951 told the story of a severely disabled man who had to rely on his fellow human beings to attend to his needs. His skull, the first of the remains to be exposed during the excavations, resembled a huge broken egg. When the rest of the skeleton was exposed, it was apparent that the man had been killed on the spot by a rockfall. His bones had been broken and crushed as a large number of rocks fell on him. His body had been thrown backward down the slight slope, while a huge boulder severed his head from his body. All this happened within seconds.

A laboratory examination of the skeleton showed him to be about forty years of age when he died, quite old for a Neanderthal. His had been a hard life. His right arm was withered and had been useless from birth, and excessive bone scar tissue on the left side of his face showed him to be blind in the left eye.

Born disabled in a hostile world, he certainly could not forage or fend for himself, so it must be assumed that he was accepted and cared for by others. He certainly could not have lived as long as he did without the care of other Neanderthals. Because he wasn't exposed to the rigors and dangers of the hunt, he probably outlived most of the people who took care of him.

His job was very likely to make himself useful around the hearth. Confined to the cold, damp recesses of the cave, he had developed progressive arthritis, which had crippled him so severely that he was scarcely able to walk.

His withered arm had been amputated above the elbow during life, and the healed bone indicated that he had survived the "operation." Possibly the withered arm was believed to be evil, and if it were severed a good one would grow in its place. As he was held down on the cave floor, the shaman amputated the arm with a blow from a stone ax.

His life finally came to an end when an earthquake caused part of the rock ceiling to collapse on his helpless body.

The excavation unearthed another Neanderthal skeleton, this one from an earlier time period, who had also been killed when part of the cave ceiling fell on him. His remains clearly reflected the hard life those humans endured—the emotion that led to his death was hate.

His skeleton lay only a few feet from the remains of a fire, on the exact spot where he had been killed by a rockfall. Under laboratory examination one of his ribs showed a strange rectangular cut. Later x-rays of the rib revealed that the cut had been made by an implement with a tapered edge. No animal attack or accident could have produced the wound—it could only have been the result of mortal combat with another human who used a weapon designed for destruction. The wound had been healing for about a week when the man was killed.

The man was one of several who lived in the cave during the climax of the Paleolithic Stone Age. For unknown reasons he had engaged in mortal combat with a neighbor and in the fight had received a nonfatal spear wound. It is reasonable to assume that he was victorious and destroyed his enemy, or he certainly would have been neither free nor able to walk away.

Returning to the cave weak and bleeding from the wound in his side, he lay down near the fire; there he remained for about a week while his wound slowly healed. Possibly asleep when an earthquake suddenly struck, he managed to get into a crouching position just as the rock ceiling fell on his helpless body.

Another skeleton of a Neanderthal man indicated quite a different human emotion—love. His remains were found deep within the recesses of the cave, where he had been buried ceremoniously by his companions, probably his family.

Studies of the encasing soil showed it to contain pollen from eight species of brightly colored flowers. No accident of nature could have caused the presence of such a concentrated botanical mixture so deep within the cave. During the harsh Stone Age one or more people must have roamed the mountainside collecting brightly colored flowers, which they then wove into the branches of a shrub. This blossom-bedecked shrub was then placed on the body of the deceased. In most modern societies it is customary to place flowers with the cherished dead, but to find flowers in a Neanderthal burial site appears quite contrary to the image of the savage caveman.

• •

When Rats Leave a Sinking Ship

It is not unusual for rats to leave a sinking ship in great columns or masses. This proverbial phenomenon has nothing whatsoever to do with any mystical powers the rat might possess. Nevertheless, the sight of hordes of rats leaving a ship might be distressing to a fellow passenger, for it is a strong indication that the ship is about to go down.

Being burrow dwellers by nature, rats live in the deepest recesses of the ship, in the bilge. This area is so low as to be almost inaccessible to the sailors. Thus the rats become aware of water entering through leaks much

sooner than the crew does. As their nesting places are flooded the rodents must perforce flee the ship. Their continuous shrill cries of alarm quickly summon the rest of their kind from the hold. They rapidly build up into a large, frightened mass of rats making a panicky exodus. This is, of course, a final calamity for the rats, as they will try to swim to eternity and usually do.

It is quite natural that a sight such as this would incite the crew to an equally harried departure, but seldom quite so disastrously.

• •

The Shadow of Death

Inhabitants of South Africa hold the black mamba, *Dendraspis polylepis*, in great dread; to them this snake is the shadow of death. Many authorities consider the black mamba the deadliest snake in the world.

Outstanding among the many old wives' tales about the black mamba is the assertion that it can keep pace with a fast-moving horse. Even though this is a gross exaggeration, these snakes can move exceptionally fast for crawling reptiles. They have been clocked at seven miles per hour and can easily overtake a slowly fleeing human. Striking swiftly, they deliver a load of venom toxic enough to kill several men. Unless medical attention is immediate, the victim will most certainly die, usually within a very few hours. Recently a man bitten on the shoulder died in less than ten minutes! The snake probably delivered the bite directly into an artery.

Being inhabitants of the dense bush and jungle, black mambas do not often come in contact with man. Therefore, despite the extreme toxicity of their venom, they are not the greatest reptilian killers of man.

The dubious honor of top killer goes to another serpent, the puff adder, *Bitis arientans*, which kills more people than all other snake species on the African continent combined! The puff adder is rather lethargic during the day, but at night it seeks out human habitations in search of rats, its chief source of food. The snake is not easily frightened, and when a human approaches it does not move out of the way and is therefore easy to step on in the dark. When this happens, the snake reacts instantaneously, and the human is bitten. The puff adder has enormous fangs and very deadly venom. Medical treatment is often not available, and usually the victim dies.

Although the puff adder is a much feared snake, it once served a useful function. Before rifles were available in central Africa, natives would capture a live puff adder and tie it to a post alongside a well-traveled buffalo trail. The furious captive snake would strike violently at the first unsuspecting buffalo that came within range. The buffalo that was bitten would soon die, thereby providing the hunters with fresh meat. And, as often as not, the snake itself was served up as dessert.

• •

Judgment Day

Although earth has existed for at least four-and-a-half billion years, it is, by astronomical standards, still relatively young. Doubtless, life on earth will go on as long as the sun continues to burn, thereby supplying the planet with life-giving energy. The sun has already been burning for over six billion years and shows no indication of ever slowing its rate of fuel consumption. Scientists believe the sun will be long-lived and that it has enough fuel for several billion years to come.

The opinion of many leading astronomers is that the ultimate destiny of our sun will be to become what is known as a red giant. Assuming that this will eventually happen, the following events will probably occur.

For aeons to come the sun will continue to burn as it does now, but eventually, as the hydrogen fuel is depleted, certain dynamic processes will cause the sun to expand. At first the rate of expansion will be relatively slow, but it will accelerate rapidly until the sun becomes a full-fledged red giant. At this stage the greatly expanded sun will fill the entire orbit of the planet Mercury and may expand even farther.

The effect on the earth will be dramatic even though it will take a billion or more years to unfold. First the polar caps will melt and the waters will cover most of the land. Temperatures on the surface of this planet will soon rise to over 1,300 degrees Fahrenheit. This will cause the oceans to boil away, and the atmosphere will be driven into outer space. The long-awaited judgment day will at last have come and gone. Life will no longer exist on the surface of the earth. But for aeons to come the lifeless, scorched planet will continue to rotate on the grill of the dying sun.

However, this intense celestial drama does not end there. Eventually, as the sun cools, it will shrink, growing smaller and smaller and shining more and more faintly. Approximately thirty billion years from the present, the last of its fires will gently flicker out. The darkness of night will then reign in our solar system for all eternity.

Perhaps the ingenuity of earth's inhabitants will prevail, and they will find ways to escape the holocaust by surviving underground or by making mass migrations to other worlds. This should be of little concern to modern humans since these catastrophic events will not even begin to happen for at least another three billion years.

• •

In the Interest of Science

Many fur trappers believe that a skunk cannot eject its scent with its feet dangling in midair. Although this opinion has been around for a long time, no one appears to have tested its validity until recently.

A dedicated scientist from a leading university decided that it was time

to prove or disprove the theory, so he experimented with a live skunk. He certainly disproved the old trappers' tales when he held the skunk by the tail in the manner prescribed: he received the full discharge of scent directly in his face! He simultaneously proved the skunk's tail isn't essential in scattering scent.

• •

Frontier Days

Bartenders with big fingers were in great demand during California's gold rush days. The man would reach into a customer's "poke" and take a pinch of gold dust, the standard price of a drink. After depositing the pinch into a box, the bartender would remove the residue by running his fingers through his beard. At the end of the day's work, in the privacy of his own home, he would carefully wash his beard over a basin. In this manner he was able to amass a sizable second income, usually much greater than his wages as a bartender!

Some remarkable people lived in California during the gold rush days, among them a stagecoach driver named Charlie Parkhurst. He smoked, drank, and chewed, as did most drivers, and during the many years he drove passengers and gold shipments along the dangerous roads, he shot and killed two highwaymen. He finally retired and went into the cattle business. In 1879 visiting neighbors found old Charlie dead at home. When his body was being dressed for burial, it was discovered that Charlie Parkhurst was a woman!

In the frontier days of the American Southwest it was common to sell to a "greenhorn" outcroppings of white rock purported to be silver. Beginning as a joke, this gradually evolved into an organized racket, until finally it boomeranged. After buying the white rock, one of the greenhorns had it chemically analyzed in San Francisco. Upon receiving the results, he quickly bought up all the greenhorn silver he could find. This fine quality of borax now bears his name, colemanite.

As an expression of incredulity almost everyone has heard the statement "Thar ain't no such animal." The speaker is probably unaware of the origin of the statement, or to what animal it refers, but knows only that it's an amusing way to suggest that seeing is not believing. Actually this crude and ungrammatical phrase has a rather comical origin in the Southwest of the late 1850s.

At that time the United States Army had imported a number of camels from Arabia to be used as beasts of burden on various desert trails in the arid Southwest, particularly in Arizona. The exact date of the following event is not on record, but it was heard by enough people to keep the expression alive.

It seems that several camels were tied to a hitching post in Tucson, Arizona, when a man, three sheets to the wind, came staggering out of a saloon and took his first look at a camel. He must have thought he was having hallucinations from too much drink, because he loudly exclaimed, "Thar ain't no such animal!" The expression took hold and is still popular today, although few actually seem to appreciate the real humor behind it.

It was on a sunny April 3 in 1860 that a young man embarked on the first leg of a 2,000-mile trek from St. Joseph, Missouri. Simultaneously in Sacramento, California, another rider began a ride in the opposite direction, and an American legend was born—the Pony Express. It was a venture that lasted only eighteen months, yet its rugged spirit typifies the West even today.

The Pony Express was started by freight magnate William H. Russell with the help of U.S. Senator William M. Gwinn. They drew up plans and mapped a trail of 157 relay stations passing through Kansas, Nebraska, Wyoming, Nevada, and California.

Prior to this new mail system, coast-to-coast mail often took as long as six months; now it was reduced to about ten days. Of the eighty or so riders in the Pony Express, half were always riding in either direction. Each man rode an average of seventy-five miles, and horses were changed every ten to twelve miles. Only two minutes were allowed to change mounts, which really left the rider with no time to rest or unwind.

The rider's trail was fraught with danger. He had to traverse unknown deserts, cross raging rivers, sweep through pathless forests, or pierce the wintry blasts of mountain passes. In stormy spring weather, the stubborn mud gripped the horse's every step. And of course there was the ever-present danger of outlaw or hostile Indians. Small wonder that recruitment posters read "Orphans preferred."

Movie and television producers have wildly embellished the story of the Pony Express riders. It seems a shame to kill the exciting stories that Hollywood has fabricated, but despite the fierce nature of the Old West, which constantly challenged the mail carrier, only one Pony Express rider was actually killed in the line of duty—and his horse, on its own, managed to get to the next station, with the mail intact!

• •

When the Gods Smile

Seven thousand years ago wild grapes were a major source of food for the people who lived along the Nile. These grapes were harvested by women and quite frequently placed in large urns. Since grapes were continuously added to those already in the urn, the original harvest on the bottom remained there for quite a while. It was soon discovered that they had a desirable quality, so they were purposely left on the bottom.

Grape skins contain natural yeast, so it was not unusual for the crushed grapes on the bottom of the urn to ferment before the top bunches were eaten. Those who were aware of the special quality of the bottom grapes would wait impatiently for the urn to empty so they could indulge in the "bubbling" nectar. The good feeling that always followed was attributed to the "Smile of the Gods."

• •

A Tomb for Khufu

Egyptians had strong beliefs in life in the hereafter. The nobility were always honored upon death, and steps were taken to provide them with every comfort in the next life. This was one of the reasons for the elaborate tomb paintings.

One of the greatest unsolved mysteries of ancient times is how the Egyptian artists performed their art in the dark recesses of a tomb. How were they able to harness adequate light for work where no sunlight could penetrate? They certainly did not use torches, because no sign of soot has ever been found on the walls or ceilings as would have remained if flaming light had been used.

Some authorities believe the answer may involve the use of mirrors. A series of highly polished metal mirrors placed at strategic angles could reflect sunlight off each other. The sunlight could then be reflected around corners and deep into the furthest recesses of the tomb.

Many of the pharaohs apparently would try to prove their greatness by, among other things, building enormous tomb structures. It is no great mystery to scholars how the largest tombs were built. What is most amazing is the creation of such magnificent feats of architectural engineering so far back in antiquity.

Probably the most outstanding example of such an achievement is the Great Pyramid of Giza. Pharaoh Khufu (Cheops) caused it to be built nearly 5,000 years ago, and it was considered to be one of the Seven Wonders of the Ancient World. The pyramid covers thirteen acres and reaches a height of 481 feet—the equivalent of a forty-story skyscraper in the modern world. Perhaps even more amazing are the building blocks of this great structure. It is made of about 2,300,000 blocks of limestone, each weighing two-and-a-half tons. Yet these gigantic blocks were fitted with no joint wider than a fiftieth of an inch!

When scientists measured the sides of the pyramid, they found that all sides differ by a maximum of only eighty-eight thousandths of one percent. And the ancients built it without the aid of precision instruments or computers; their device for these incredibly accurate measurements was a knotted string.

For the modern reader of history it is difficult to comprehend how ancient the great pyramids at Giza really are. Well over two thousand years ago, when the Greek historian Herodotus visited these great wonders of the ancient world near modern Cairo, he was, like any other traveler, looking at an ancient tourist attraction. Even at that time the pyramids were already over 2,000 years old!

In contrast to their present-day rugged appearance, the Egyptian pyramids were once completely faced with highly polished marble. When Herodotus traveled in Egypt and later described the pyramids, his writing emphasized the marble facing, which was so highly polished that he could see the clear reflection of the clouds passing overhead.

The pyramids today stand relatively intact despite nearly fifty centuries of wars, earthquakes, tourists, tomb robbers, and natural decay. They have come to symbolize eternity itself.

Hurricane Celebration!

Often people experiencing hurricane fury feel that the earth appears to be in its death throes. Hurricane Camille created such an effect.

On August 17, 1969, Camille roared across the Gulf Coast, slamming inland near Gulfport, Mississippi. Winds at speeds of over 200 miles per hour lashed the earth, and a deluge of rain fell on the land. When its fury was abated, Camille left behind over 250 dead and about eighty missing. The property damage amounted to well over $1.5 billion.

When Hurricane Camille first came roaring out of the gulf and rode over Mississippi, some 75,000 people fled inland to higher ground. However, a few of them decided not to join the timid escapees.

In the town of Pass Christian, Mississippi, twenty-five guests of the resort hotel Richelieu made a fateful decision. Camille was already thrashing at the hotel when the twenty-five fun lovers decided to throw a hurricane party! And so, as officials pounded on doors shouting last warnings to evacuate, the guests, amused at their own trickery, kept very quiet. Convinced that the hotel was completely evacuated, the officials withdrew and sought safety inland.

The hurricane party began but was rather short-lived. The storm completely demolished the hotel, killing twenty-three of the party group. The two survivors would make only one brilliant comment to justify their actions: "We thought it would be great fun!"

• •

The Deadly Female

Mating can be fatal for the males of some species. For the preservation of the race the male must occasionally sacrifice himself to the deadly female.

Female scorpions are guilty of the most atrocious table manners, and their invited guests, their mates, often become their meals.

Scorpions are distinctly antisocial and usually keep to themselves. But when a male and female meet for courtship and mating, the reclusive

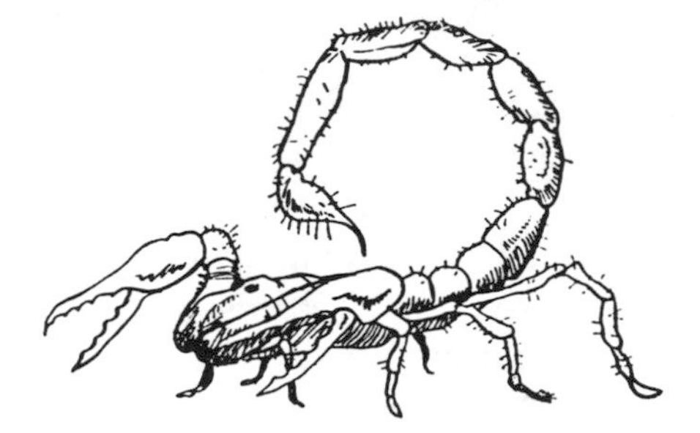

behavior becomes transformed into an idyllic relationship filled with graceful acrobatics. They begin by locking pincers and going through a strange rotating motion that resembles primitive dancing. The dance goes backward and forward, with the female following the male's lead in perfect rhythm.

The honeymoon lasts only a short time, however, and only one of the lovers survives the marriage bed; the survivor is always the female. Following the actual mating, she stings the male to death and consumes him.

The female praying mantis does not consider the mating act complete until she has killed and eaten her mate. So cannibalistic is she that her mate is often killed during the courtship ritual. Not infrequently, while mating is occurring, she turns and clamps her forelegs in a viselike grip on the male and kills him by biting off his head. It is therefore not unusual for a female praying mantis to have more than one mate within a single season. One female was observed to have devoured eight of her suitors.

The most famous of the deadly females is, of course, the black widow

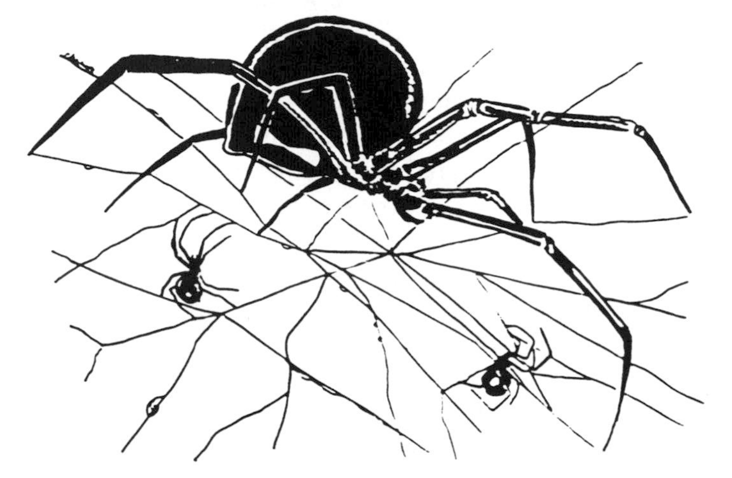

spider. Very little is heard of the male of the species. He is just as abundant as the female, but only one-sixth her size. Perhaps the male black widow's greatest claim to fame is that he has never been known to bite humans.

Despite her reputation, the female black widow spider consumes her mate only if she is hungry. During the courtship period she is as tender, loving, and sensuous as could be expected of a female spider. The male is instinctively aware of his danger, so while she is in a passive mood he often ties her legs with web silk to immobilize her and then proceeds with the business at hand.

Following the actual mating the female easily breaks her bonds. During the brief moment while she is freeing herself, the male has time to make his getaway—but not always.

• •

"In the Bloom of Youth"

The crystalline beauty of snow and ice often completely masks a most dangerous and dramatic force of nature. In the European Alps alone over 10,000 avalanches occur per year. The death toll during this century numbers well into the thousands.

The sight of mountain climbers in the Swiss Alps, with guides who appear so expert in their craft and surefooted in the snow, is impressive. But even the advantage of much experience and knowledge of the ways of snow and ice is no guard against sudden unexpected disaster. It is almost as if nature is luring these adventurers into a catastrophe from which there is no escape.

A rather famous avalanche story with a grisly ending occurred on August 20, 1820. On that occasion a nine-man team was attempting to climb Mont Blanc. The expedition was going along without mishap when suddenly, with no warning, the icy snow above cracked and came down on them. An enormous avalanche swept three of the guides into a glacial crevasse, where they perished, eventually to be assimilated by the glacier. By some miracle the remaining members of the group were unharmed and returned to report the deaths to the authorities.

Because they knew the location of the bodies and could calculate the rate of glacial movement, they predicted that the bodies encased in the ice would reappear forty years later at the foot of the valley of Chamonix some five miles away. Their calculations proved to be surprisingly accurate, for the bodies of the three guides did appear, completely encased in ice, but forty-one years later. According to reports of witnesses, "They were still looking in the bloom of youth."

Although a climber has no recourse against an avalanche, such events do not always end in complete tragedy. In one case it appears that the hand of Providence tried to intercept the angel of death.

It happened in 1952, when a group of Swiss mountaineers were horrified to look up and see an enormous avalanche plunging straight toward them. No escape seemed possible, and they expected to be crushed and buried, when suddenly a large crevasse opened up between them and impending death. The mass of snow and ice crashed in its entirety into the opening, leaving the men terrified but completely unharmed. The sudden opening was a collapsed snow bridge over a large glacial crevasse. Ironically they had walked over the bridge just moments before, completely unaware of the thin covering between them and yawning death below. The men were so emotionally distraught by the experience that one of them never really recovered and eventually shot himself.

A cruel fact about avalanches is that they occasionally repeat long after the apparent danger has passed. After survivors of a snowslide were rescued in Blon, Austria, in 1954, a second slide occurred without warning and reburied the twenty-five survivors along with their rescuers. A seventy-year-old woman was recovered alive after a fifty-hour reburial in the same snowslide.

• •

The Trail of Tears

One of the most tragic chapters in the history of the American Indian came about as the result of the United States government's plan to relocate the Cherokee Indians to a reservation near the present city of Tulsa, Oklahoma.

The series of events that led to the tragedy began in 1827. In that year the Cherokee boldly declared their nation to be an independent republic and even ratified their own constitution. In 1828 a renowned Indian fighter by the name of Andrew Jackson frowned on recognizing the Cherokee as a nation, let alone a republic. When gold was discovered on Cherokee land the following year, the Jackson administration enacted the Indian Removal Act, which gave the United States the power to decree removal of Indians from any designated area. The Cherokee were now legally required to leave their land and homes and move to what was, to them, a foreign country, Oklahoma Territory.

The law was not really enforced for several years, but some of the Indians, highly embittered, voluntarily emigrated from the area. In 1838 the U.S. Army moved to forcibly evacuate the remaining Indians from their homeland, which made up most of the state of Georgia.

The army officer in charge of the removal of the Cherokee issued precise, strict orders that the evacuation be carried out as humanely as possible. His men were apparently less sympathetic than their commanding officer, for in every operation they completely ignored his orders for compassionate treatment of the Indians. Cherokee families were virtually herded out of their houses at bayonet point, which often "accidentally" inflicted painful wounds and resulted in several fatalities. Men, women, and children were shoved, beaten, and even shot, while bands of civilian looters pillaged Indian villages, sometimes before all the residents had left. Not even the cemeteries were spared as savage brigands unearthed graves, searching for silver pendants and jewelry. The skeletal remains were ruthlessly and contemptuously scattered in all directions.

The Indians were at first interned in makeshift camps and then grouped into caravans. In September 1838 they left Georgia and began the long trek to their new homes in Oklahoma Territory.

The westward movement was very slow; some rode in crude wagons and some on horseback, but by far the majority were on foot. They trudged painfully through an unusually snowy, cold winter. The price was high for the old and feeble, who withstood as much as flesh and blood could endure and then began to die. Travelers who passed the processions all described the constant wailing and weeping as the caravan stopped momentarily to bury the dead. The survivors, their faces wet with tears—very unusual for Indians of that time—left their loved ones behind. They genuinely felt that those who had died were the fortunate ones, and this knowledge somehow relieved their grief.

The last of the Cherokee arrived on the reservation near present-day Tulsa in March 1839. It is not known exactly how many had perished along the way, but a conservative estimate is 4,000—nearly a third of the Cherokee population. Superstition soon set in, and tales still abound. Travelers near the route of the caravans vow they clearly hear the wailings of women as the gentle winds blow in the night.

By 1840 the route of the Cherokee became known as "The Trail of Tears," a title that persists today.

• •

Tools of the Times

Sea otters spend almost all of their time in the ocean, in coastal waters that provide them a steady diet of sea urchins, clams, crabs, snails, mussels, and particularly abalone. They are most at home amid kelp and seaweed, where they can become entwined in kelp to avoid drifting off while they sleep.

They scour the sea floor in search of prized delicacies, and when they find a mollusk or an urchin they swim to the surface, lie on their backs, and crack it open by pounding it with rocks. This may now be considered old-fashioned, because at least one sea otter has adapted to the twentieth century. It was observed by a scientist recently in the process of opening a shellfish. But this sea otter had cracked open its abalone shell with a Coke bottle!

• •

CHAPTER TWO

In Solitude He Wanders

Most people consider the lion by far the most dangerous animal in Africa. There would be much debate among professional hunters on this subject, because those who have had contact with creatures in the wild consider the buffalo the most dangerous among big-game animals.

If lions could express their opinions they would probably agree. A recent

well-recorded incident tells about a herd of buffalo that passed an attacking lioness among them on their horns! When they finally relented and the lioness was allowed to drop to the ground, she was dead.

Lions have a definite social order, with one male in a dominant position. Although a large portion of the pride's activities are attributed to the female, she is not alone during times of dire need.

When a lioness senses she is about to give birth, she will isolate herself from the pride. Before retiring, however, she finds another female to assist in her confinement. The task of her companion is to protect mother and newborn cubs from predators and to provide them with fresh food. The Masai tribesmen refer to the assisting lioness as "Auntie."

As soon as the mother regains her strength, Auntie leaves and returns to the pride. The mother, after hiding her cubs, also resumes hunting with the pride. She is, as often as not, a devoted mother. Recently a park ranger observed a lioness drag a 300-pound sable antelope almost a mile to her cubs—a tremendous feat of strength and endurance.

When the cubs are three months old, the lioness introduces them to the adults of the pride, where they will be accepted as members of the band. If this were not done, the cubs would be killed on sight as intruders, because lions of different prides are quite hostile to each other.

In any pride of lions a definite hierarchy exists. When one of the group makes a kill, the dominant male is the first to grab a choice portion. He is followed by the mature females and then the juveniles—that is, provided they can find an open spot, as by now the carcass is fairly well overrun with feeding lions. In such a situation the cubs are left out. They usually try to join in, but are often actively restrained from joining the feast by their mother. She is well aware that the defenseless intruding cubs would be instantly killed or mauled. It is small wonder that only one cub out of ten survives to become an adult.

The new mother's mate is the ultimate of male chauvinists. He leaves all the major necessities for living, such as hunting and killing, to the female. Indolent and lazy, he spends as much as twenty out of twenty-four hours just resting. He seldom arouses himself except to eat, but when he does, the male lion gorges himself on as much as ninety pounds of meat at a single meal! But even the king has his day of reckoning.

In the social order of the king of beasts there comes a time when an up-and-coming young male lion decides it is the hour to dethrone the boss lion of the pride. He will approach the boss somewhat cautiously and then force matters into a fight. It begins with bluff as both males try to howl each other down. The sounds they emit range from low growls to nerve-shattering roars that can be heard for miles. Soon the blows begin to fall.

Each lion beleaguers the other with slaps of a force that could easily break a man's neck. However, the male lion possesses a well-padded mane that, in this instance, becomes a shock absorber. During the fight neither

exposes his claws, so the fight involves mainly muscular power and is really a fair and bloodless contest for superiority.

If the boss lion wins, nothing will change, and the defeated challenger will sulk back to the pride. Conditions will return to normal as neither harbors a grudge.

If the challenger wins, however, everything changes. Most, if not all, of the pride members will then fall under the jurisdiction of the new boss lion. The now dethroned male is sent out "into the desert." From here on he will roam as a "solitary." Since the territories of neighboring prides of lions always border one another, the fate of the former boss lion is harsh. If he tries to enter the realm of another pride, he will immediately be chased away or killed.

The solitary lion is constantly harried and forced to move on and on until he reaches a parched land that lions do not frequent. Here he must live on rodents and lizards, for the larger game do not try to subsist in this type of environment. Worse still, he may be forced into the vicinity of human settlements, which to a lion is extremely distasteful. He will now develop skills in hunting domestic animals, thereby becoming a rogue who is constantly threatened by the rifles of the humans. In such a situation it is not unusual for the lion to become a man-eater. Exiles such as the solitary lion will, in sheer desperation, attack man. Once this happens he becomes a confirmed man-eater and is hunted down in earnest—finis.

The Amazon Molly

There exists in Central America a race of female fish that survives almost entirely without males. Their lifestyle recalls the legendary Amazons of Ancient Greece—women so strong that only the most skillful of male warriors could hold their own against them.

The Amazon molly, *Mollienisia*, is closely related to the guppy and is no larger than a man's finger; being a race of female fish, they never give birth to males! Since there are no males to father the species, it is necessary to

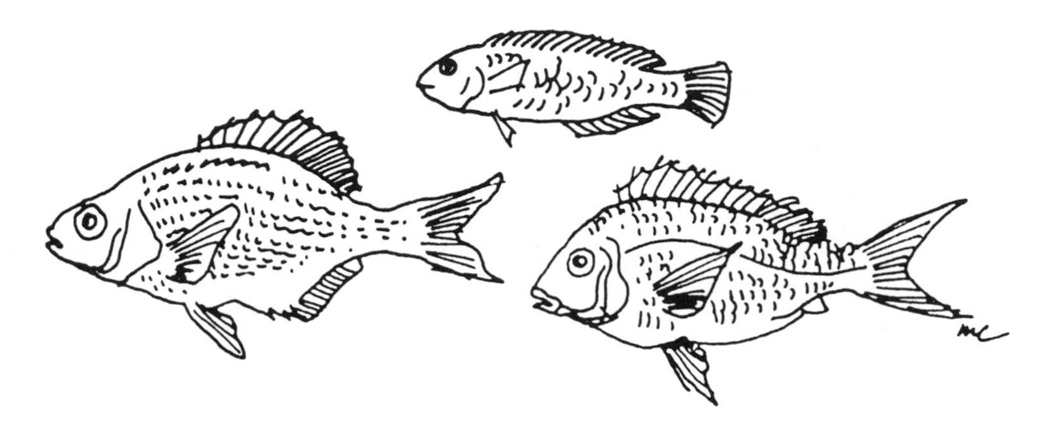

"borrow" males of a closely related species for purposes of conception. This happens only once a year, as it did among the Amazons of Greek legend.

The sperm of the male does not actually fertilize the eggs of the molly. It does penetrate the egg cell, but only to activate cell division. By this process a single cell will develop into the billions of cells that compose a living creature. As soon as the process is initiated, the sperm degenerates without fusing with the egg nucleus.

Because of the degeneration of the sperm, none of the hereditary traits contained in the sperm can possibly be inherited by the young. This is quite different from the young of most other species, for when the sperm fuses with the egg, the offspring will inherit traits from both parents. In the case of the Amazon molly it could be said the male chromosomes are "murdered" in the female cell. This also recalls the legend of the Homeric Amazon women, who routinely destroyed any male offspring.

The method of reproduction of the Amazon molly is a very specialized form of parthenogenesis—birth without fatherhood—known scientifically as gynogenesis. A strange brood of young is produced as a result of this unusual conception. Born into the world are only females, identical down to the last scale, each an exact clone of the mother. All have inherited exactly the same traits since there has been no mixture of genes that they could have inherited from the father. Down through the millennia, each generation has remained identical to all those that preceded it.

The lifestyle of the Amazon molly clearly shows what human life would be like if both sexes were not actively involved. The fish have no individuality. This race of females is not a primitive form of fish, as might easily be concluded; on the contrary, they are quite an advanced form of life, highly specialized in order to preserve their race. As stated above, they are closely related to the guppy, a family of fish that gives birth to its young alive. All males in this family are cannibalistic. When a female starts to give birth, the male swims under the female swallowing the young as rapidly as they are born.

Perhaps the cannibalism of the male guppy serves as a rather barbarous method of birth control, thereby preventing overpopulation and its consequences. Nevertheless, the female aims to prevent the loss of her brood and must find a hiding place to produce her young. This becomes rather difficult when the fish population of an area is high, which it often is.

The threat of cannibalism has been overcome by the Amazon molly. They have their own territory and leave it only temporarily to mate with the selected species of males. This completed, the female will return to her territory. If the male tries to follow, she will ram and beat him with her fins until he is driven away. As a result, no male ever enters the realm of the Amazon molly.

The Amazon molly is certainly a specialized form of life that has achieved an ultimate adaptation in protecting the young from the cannibalistic male guppy. Since this illustrates how an animal society can exist without males, one cannot help but reflect—did the legendary Amazon female of Homer's day actually exist?

• •

Bird Brains

At times woodpeckers seem to be overzealous in storing their winter food. They will hammer out holes in a tree with their picklike beaks and stuff in nuts and acorns. It is not unusual for them to include stones and small pebbles indiscriminately as winter "food."

Sometimes a particularly vigorous woodpecker, working on a thin softwood tree, manages to bore a hole completely through the tree. But apparently it doesn't notice; in goes acorn after acorn, nut after nut, all of which fall uselessly to the ground on the other side of the tree. Handfuls of acorns and nuts have often been found lying on the ground under woodpecker holes. The birds that do this probably starve to death during the long and harsh winter.

The hoatzin, a bird the size of a pigeon, lives in the swampy jungles of the Amazon valley, always close to water. Although it can fly, it rarely does; instead it spends most of its time in water and has evolved into an excellent swimmer. Its nest is built in low branches only a few feet above the surface of the water. At the approach of an intruder, young and old alike dive headfirst into the water. This is all right, except that, as often as not, the gaping mouth of a crocodile waits directly under the hoatzin nest. But this seldom deters the not-too-bright bird, who dives anyway—right into eternity.

The turkey, symbol of the American Thanksgiving, is not a notably intelligent bird, particularly when it is young. Newborn chicks have to be taught to eat, or they will starve. Breeders spread feed directly underfoot, hoping they will get the idea. During a heavy rainstorm the young turkeys look up with their mouths agape; a few have actually been known to drown!

There is another type of "birdbrain"—not really stupid; rather just plain crazy. During the middle of October every year, for completely unknown reasons, some ruffed grouse engage in very abnormal bird behavior.

During what has become known as "crazy season" many of the grouse are found wandering aimlessly and apparently in a dazed condition up and down village streets of the Midwest. Several will wander into barnyards, mingle with, and act just like domestic chickens. Others will fly at high speed into open windows and flap aimlessly around the room, overturning flowerpots and smashing chinaware. Still others appear to kill themselves deliberately by flying against stone walls or other solid objects. A few of the crazed grouse become as tame as kittens and stroll up to some human to have their head scratched or petted; these often wind up on the dinner table.

As yet no accepted scientific explanation has been formalized, but it has been suggested that the fermented fruit of some species of plant that ripens at that time of year is catalyst for their unorthodox behavior.

Perhaps it is fitting to end this dissertation of not-too-bright birds by briefly describing a bird that, by human standards, is completely without modesty. The bird is the South American tarpory, which spends most of its day pecking around the ground for food. This is not too unusual in itself, except that the bird appears to have an extraordinary digestive system and excretes more or less continuously while it is feeding. Since it is almost constantly on the ground feeding, the tarpory frequently leaves behind itself a twenty-foot trailing ribbon of dung. The tarpory thus becomes easy prey for predators, which simply follow the dung trail. Easy pickings for the hungry carnivores, because the bird is usually too busy eating to notice that it is about to become dinner itself.

• •

Death from Outer Space

For a human or an animal to be struck by a falling meteorite is extremely rare, and yet several incidents have been recorded. A leading scientist lists twenty such events, rating them from certain to dubious. The list includes meteorites from prehistoric times to the 1950s.

The entire list includes only five recorded as certain. The most famous is the Sylacauga, Alabama, meteorite of November 30, 1954. The object first struck the roof of a house, fragmenting on impact, and then tore into the house. Some of the fragments struck a woman occupant of the house on her arms.

On July 14, 1847, a forty-pound meteorite struck a house in Braunau, Austria. It also fragmented upon impact and penetrated the house. Several fragments fell on a bed where three children were sleeping, causing more alarm than harm.

One of the incidents listed as uncertain took place in prehistoric time. Despite its rating, there is strong evidence to indicate it probably did happen. This prehistoric entry involved a stony-iron meteorite and a *Megatherium*, an extinct giant South American sloth.

The broken skeleton of this Ice Age giant was found lying on the meteorite. Although the meteorite could have fallen ages before, it would have been an amazing coincidence for the sloth to have toppled over on it as it died. Moreover, the nature of the skeletal remains has led many scientists to suspect that the sloth was struck by the falling meteorite—some 20,000 years ago.

• •

The Sound of Music

In 1964, a geologist and an archaeologist were exploring several of the shallow limestone caves in southeastern Arizona. They found a number of fossils and Indian artifacts at various localities but finally came upon the unexpected.

In one of the caves they found an Indian mummy sitting cross-legged in his tomb. The skin of his face had shrunk, resulting in a perennial death-mask grin. It was almost as though he were greeting the scientists with a toothy smile. He was naturally nicknamed "Smiley."

Along with Smiley were various articles that he had owned in life and that were left with him for use in another world. They consisted largely of baskets, sandals, bits of flint weapons, and, as a most unusual accessory, a carved flute that rested on his lap. He must have treasured this musical instrument in life and undoubtedly had fashioned it himself.

Wondering if the flute was still workable, one of the scientists reached down and picked up the instrument. Unable to resist the temptation, he raised it to his lips. Across the rugged landscape clear musical notes rose into the pure air in a sound that had not been heard for over 2,000 years.

• •

Rhino!

Weighing more than a ton, standing five feet tall and twelve feet long, the rhinoceros looks every bit the prehistoric monster that it is. A leftover from the past, the rhino has changed very little from the time its ancestors roamed the earth in the company of mastodons and saber-toothed cats.

The African black rhinoceros was a common sight fifty years ago, but now fewer than 15,000 survive, and their numbers are constantly dwindling. Rhinos have not become evolutionary failures; their undoing has been rather the lancelike horns jutting from their heads. All five species of rhino are being hunted into extinction because of the widespread belief throughout Asia that ground-up rhino horn is a powerful aphrodisiac. Because of its reputation as a sexual stimulant, rhino horn has become the most valuable animal product in the world. A pound of powdered horn is worth its weight in gold in the Orient, where a single horn can bring several thousand dollars. As a result a strictly illegal business has flourished while the rhino population diminishes.

It is unfortunate that in the mind of the user the ground-up substance added to a potion he drinks will send him into a sexual frenzy; in reality rhino horn is no more of an aphrodisiac than is an ordinary lump of sugar!

The three Asian rhinoceroses are almost extinct; concentration has therefore turned to the African rhino, which, like its Asian cousins, is now becoming a victim of the quest for sexual prowess.

The black African rhino has a vile temper, which makes it extremely unpredictable. Perhaps its poor eyesight is the factor that makes it defensive and untrusting and therefore irritable. It is impossible to know whether the rhino will ignore an approaching human or charge without warning or provocation. This may serve as a protective mechanism, as doubtless many a poacher has paid with his life when the rhino suddenly turned on him.

The irrational behavior of the black rhino does, at times, reach ridiculous

proportions. A game warden in Kenya tells how he pulled a rhino out of a mud hole with his Land Rover. The beast promptly repaid the favor by caving in the side of the vehicle. It was almost as though he blamed the vehicle for his having gotten stuck in the mud in the first place.

Lately the rhinos seem to have been on a rampage to reduce the African population of automobiles. Reports of rhinos blindly charging automobiles and keeping up the attack until the vehicles are reduced to wreckage have increased. Some have adopted an even more effective method in the one-sided war against autos. They stand near the highway, apparently waiting for a car to approach within range, and then step out into the road. For a small speeding car, a collision with a full-grown rhinoceros is like crashing into a stone wall.

One rhino attempted the ultimate conquest. In a well-recorded incident a large male rhino tried to take on a railroad train! Passengers were rudely made aware of the attack by a tremendous jolt. Looking out of the window, they watched as the rhino scrambled to his feet and staggered off, snorting and shaking his head. He emerged the loser of that fight but got off with nothing more than a monster-sized rhino headache.

It seems rather inconsistent that an animal with such a vile, unpredictable temper is considered the most easily tamed animal in Africa. In captivity the rhino becomes so gentle that it will eat out of its keeper's hand, come on call to have its ears rubbed, and may even roll over on its back so the keeper can scratch its stomach. Such activity is not without hazards to the person attending to the rhino's whims. When a rhino's ears are rubbed, it often displays its affection by trying to lean fondly against the scratcher. The animal may be only expressing contentment, but two tons of animal bulk leaning against a 150-pound human can be, to say the least, a rather unnerving experience.

• •

The Migration of the Monarchs

The monarch butterfly ranks among the greatest wanderers of all time. When preparing for a migration, these butterflies collect in tremendous swarms until the air seems filled with butterflies.

Throughout the summer, when the flowers are in bloom in southern Canada and the northern United States, the monarchs flit gracefully about as solitary individuals in gardens. But in September they congregate by the hundreds of thousands. The growing hordes become so great that tree branches actually bend under their weight. The insects are assembling for their migration southward, one of the most extensive and remarkable mass movements of any living creature.

The migration starts late in September. The butterflies travel en masse, spanning a continent and covering a distance of, in some cases, nearly 3,000 miles. In migration, the monarchs generally fly at heights of about 300 feet, at cruising speeds of about eleven miles per hour. Flying doggedly through all kinds of weather, these fragile insects, whose weight is about that of a tiny bird feather, may travel as far as eighty miles in a single day.

A rather amazing feature of the autumn migration is that the butterflies always gather on certain trees along the way. These special "butterfly trees" are hosts to the monarchs year after year, serving as an overnight stopover. Why they select the same trees every year, so far from the point of origin, is as yet unknown.

The monarchs fly as if guided by a compass; if they meet a solid obstruction, they rise above it, never changing direction as they wing toward their goal. As they are joined by several flights along their journey, they seem to swell to unbelievable numbers. One of the greatest witnessed migrations occurred over Texas in 1921.

At that time an immense swarm of monarchs, well over 250 miles long, passed over the Lone Star State. Scientists estimated that at any given sector along the migration route at least one million butterflies passed during every minute of daylight—and this lasted for about eighteen days!

As they reach their final destinations, the butterflies gather in tremendous concentrations along the Gulf Coast and in California from Pacific Grove along Monterey Bay southward nearly to Los Angeles.

The butterfly flocks seem to prefer Pacific Grove in California; they arrive there late in October by the millions and cover acres of pine trees. Pacific Grove is proud to be host to the monarchs and has become one of the very few insect sanctuaries in the world each year. The butterflies attract thousands of tourists who come from all over the world to see the "butterfly trees," and indeed it makes a spectacular sight. Street signs direct visitors to the butterfly centers. To celebrate the monarchs' return, schoolchildren stage colorful parades annually. In honor of their insect guests, the seal of the Pacific Grove Chamber of Commerce depicts a monarch, while a city ordinance makes it a crime to harm a monarch butterfly.

However famous the Pacific Grove sanctuary, monarchs *do* winter elsewhere, particularly along the Gulf Coast from Texas to the swamps of Louisiana and parts of Florida. Some travel from British Columbia to the Gulf of Mexico to escape the cold, but nowhere is there such concentration as in Pacific Grove.

There are times when the great hordes of butterflies do not reach their destination. They have been observed by mariners well over a thousand miles at sea. They were undoubtedly blown off course by strong winds and were winging their way to butterfly heaven.

In the spring the monarchs leave almost as suddenly as they arrived—not as successive waves of migrants but as individuals that will face the long trip north alone. The survivors arrive on tattered wings and with faded color and lay their eggs on milkweed alone. This is the sole diet of the caterpillars. In about two weeks the caterpillar encases itself in a pupa, and in about another two weeks a monarch butterfly emerges.

With the arrival of next September the new generations of butterflies will assemble as did their predecessors, and once again the sky will be brightened as the hordes of insects fill the air on their journey south. This cycle has gone on for many centuries and will doubtless continue for many ages yet to come.

• •

The Ancient Sport of Kings

The domain of the lion once extended over much of the world; its most formidable enemy has always been man. Egyptian, Assyrian, and Persian monarchs considered it their royal duty to wage continuous war against the lions. They often sought renown as lion killers under the pretext that they were protecting the peasants.

It is not uncommon for Assyrian relief sculptures to depict royal lion hunts. They usually show the lion displaying magnificent strength and courage, and in several sculptures it is shown attacking the king's chariot while the king, with equal courage, is calmly firing arrows at his opponent.

The royal hunts of ancient Asia Minor later became ceremonial rather than actual hunts. The animal, which was properly drugged, was released from a cage into a hollow square formed by troops with shields. The purpose of the troops was to position the lion so the king could kill it with little risk or difficulty. Since the lion was already befogged, it usually didn't even see the king who wielded the executioner's spear. The celebrations of the evening were lavish; after all, the king, at great personal risk, had proven his eminence by "single-handedly" killing the greatest of beasts.

• •

"The Sea Is Angry"

Oceans have always stirred to the passage of the wind. Storms cause immense waves at sea. To the mariner this is a time when "the sea is angry."

Despite the fury at sea, it is around the shorelines that storm waves cause the most destruction. Thundering breakers may shatter buildings, engulf lighthouses, and destroy the most massive shoreline installations as if they were nothing but miniature models. The force of a wave during a winter gale may be as much as 6,000 pounds to the square foot.

Consider the events at the breakwater near Wick, on the northeastern tip of Scotland, in 1872. Storm waves tore loose a concrete pier and moved the entire structure, weighing more than 1,350 tons (2.7 million pounds), inland. And this was a mere dress rehearsal; the new pier, which weighed well over 2,600 tons, suffered the same fate five years later. It was as though the sea was playing with a new toy.

Lighthouse keepers, exposed to the full strength of the surf during a storm, witness happenings so unusual that they appear to border on the supernatural.

One such incident occurred in 1840 during a heavy storm, when the heavily bolted door of the Eddystone Lighthouse in England was broken

from within, and all its iron bolts and hinges were torn loose. Scientists attribute such an event to pneumatic action. This is created by the sudden back draft when a heavy wave recedes, combined with an abrupt release of pressure on the outside of the door. Such forces undoubtedly broke the entrance door of the Eddystone Lighthouse but to the keeper they must have appeared beyond belief, perhaps supernatural.

Early in this century the keeper of the Trinidad Head Lighthouse on the coast of Oregon observed a gigantic wave rise as a solid wall of water, breaking over the top of the 196-foot-high tower. The shock of the blow caused the light to stop revolving.

Along rocky shores storm waves are often armed with rock fragments. The windows of a lighthouse on the summit of a 300-foot cliff at Pentland Firth, Scotland, are repeatedly broken by wave-tossed rocks. Once a rock weighing 135 pounds was hurled over a lighthouse one hundred feet above sea level. In falling it tore a twenty-foot hole in the roof.

As one might expect, most deaths during hurricanes are caused by violent storm waves. The most severe destruction by storm waves ever recorded took place in the Bay of Bengal on October 7, 1937. Over 300,000 people were killed, and well over 20,000 boats were destroyed.

• •

Female Courage and Pride

To the casual observer a herd of feeding elephants is oblivious to the world around them. In a way this is true, because their concentration is on feeding. They are far from defenseless, however, because feeding elephants always post sentries at critical points to warn them of an approaching enemy, usually in the form of man.

When danger does threaten, the sentry raises its trunk, and the herd is instantly alerted as far as a half mile away. The method of communication is not yet understood, but many scientists believe that a sound inaudible to humans is emitted by the elephant sentry.

Among the elephants it is usually the female who has the courage to face

whatever peril is threatening, and the male elephant is the first to flee. The female will get between the male and the source of danger and actually push him into the dense brush out of harm's way. Usually the brave male keeps right on going, leaving the female and her youngsters to get out of trouble the best way they can.

Merely wounding an elephant can be quite dangerous, whatever the nature of the wound. An incident in 1980 involved an accident between a motor vehicle and an elephant. An African game warden driving his field car on a dirt road swung around a steep curve after nightfall and unwittingly collided with a departing female elephant, striking her on the hind legs. In retaliation the offended pachyderm promptly sat down on the hood of the car, completely crushing it and causing both front tires to blow out.

Apparently satisfied, and none the worse for wear, the elephant got to her feet and casually wandered off. The car did not fare so well; it was a total wreck, and the warden had to signal for help. Fortunately the radio was still working.

• •

Ishi—Last of the Yahi

It was early morning on August 29, 1911, in Oroville, California, about seventy miles north of Sacramento. The excited barking of dogs awakened the ranchers, who, upon investigating the disturbance, found a man leaning against the fence. He was obviously an Indian, with his face blackened and hair singed to show he was in mourning. Although no one knew at the time, he was the last living member of an Indian tribe known as the Yahi.

His tribe had been a true casualty of the California gold rush days. In 1865 the Yahi had been attacked by a party of whites who murdered virtually every man, woman, and child they could find. Only four Indians survived the massacre. For almost the next fifty years they hid in the shadow of the expanding white civilization. Not a trace of their presence was ever found, and the Yahi were considered extinct.

The lone surviving Indian was first turned over to the sheriff, who called in anthropologists from the University of California, Berkeley. Eventually, as the scientists managed to communicate with him, they pieced his story together. He was in truth the last living member of his tribe. He had been totally alone for three years when, unable to find food, he finally surrendered himself to the whites, fully expecting to be killed. What was most amazing to the scientists was the realization that he had lived in the California wilds for almost half a century within a stone's throw of civilization without in any manner betraying his presence.

He was moved to the Berkeley museum, where he was employed as janitor, a position he maintained until his death five years later. He performed his duties very well, learned about 500 words of English, and quickly became a part of twentieth-century civilization. He was befriended by all who came in contact with him and even received a marriage proposal. He often toured San Francisco, where he enjoyed riding the famous cable cars. He made several trips back into the wilderness to show his benefactors what life was like in the "Stone Age."

The Yahi were forbidden ever to utter their own names, and all who knew his name were dead. The scientists therefore adopted a word from his language and called him Ishi—meaning "man." On March 15, 1916, the last wild Indian of North America died of tuberculosis. Ishi had finally returned to sit by the council fire of his tribe forever.

Within the scientific community there has been much speculation on the possibility of unknown humans still existing on this continent. Most scientists, however, while they recognize the possibility, believe that Ishi was indeed the last "Stone Age" American.

• •

Cleopatra's Asp

It biteth like a serpent and stingeth like an adder.—Proverbs 23:32

The asp *Vipera aspis* is a venomous snake common in many parts of Europe, living mostly in hilly or low mountainous country. It has been recorded at 9,700 feet in the Alps. The snake is generally slow-moving but aggressive and has been responsible for a few fatal bites, especially in southern France.

The asp is best known as the snake Cleopatra used to kill herself. Yet it is very unlikely that she would have used the asp *Vipera aspis* because this species does not live in Egypt and never has. There is good reason for Cleopatra's snake being called an asp: the name *asp* used to be a general term for any kind of venomous snake, in much the same way that *serpent* is used to describe any snake.

For thousands of years it was believed that Cleopatra committed suicide by letting an asp bite her. The drawback to using a member of the viper family for suicide is that their bites are not often fatal; moreover, the effects are painful and messy. On the other hand, the biting cobra injects a venom that interferes with the action of nerves and muscles. Death is quick and relatively painless. Scientists and scholars believe the snake that caused the death of Cleopatra was actually the Egyptian cobra, *Naja haje*.

Statues, paintings, and various other forms of Egyptian art often show

royalty wearing headdresses or amulets depicting the cobra. They may have even worshiped this snake. It has always been popular with Egyptians; it was both feared and respected. This is natural, since the snakes were very abundant and lived in close proximity to people. As a matter of fact, even now it is the only cobra in Africa known to kill a substantial number of human beings annually.

The Egyptian cobra feeds mostly on rodents, which are abundantly associated with man. The hunt for rats and mice often takes the cobra into human dwellings. Sometimes after a meal the cobra remains in the house digesting its food. In a leading hotel in Cairo a female tourist sat down one evening on a comfortable couch to read a magazine. A cobra that was resting on the couch must have panicked when the tourist sat on it. The bite she received proved almost fatal for the surprised woman. Quick medication doubtless saved her life. Cleopatra was less fortunate.

In ancient Egypt the cobra was often used as an instrument of mercy for condemned political prisoners. Since its poison was known to be quick-acting and almost painless, it was at times offered as an alternative to a more painful and dishonorable way of dying. On record are many cases of Egyptian noblemen and royalty who deliberately went to their reward via the cobra route. As scholars argue, why should Cleopatra have been any different? A bite from the Egyptian cobra brought a rapid death far more merciful than she could expect from the advancing Roman conquerors. And most important, at least to her, suicide by cobra bite was an honorable death. It is really quite doubtful that Cleopatra ever heard of an asp!

• •

Inside a Killer Tornado

Nature's most violent wind is the tornado, a vortex whirling up to 300 miles per hour, possibly even more! When a tornado is forming, a funnel cloud takes shape in the center of massive black clouds and bores downward, twisted and whirling. According to some observers, it resembles an elephant's trunk or a huge dangling rope.

Destruction begins when the funnel bites the ground with a terrific roar described as "the noise of a thousand trains" or "the buzzing of a million bees."

Tornadoes occur worldwide, but scientists agree that these destructive storms are most prevalent in the United States. This is an understated observation, since 90 percent of the world's tornadoes occur in this country, with the most common setting being the Midwest. There tornadoes kill or injure scores of people annually and cause property damage in the millions of dollars.

The damage a tornado can inflict on a city staggers the imagination. Damage results not only from the extremely high winds, but also from the intense low pressure at the storm's vortex. When a tornado strikes a building, the low pressure within the vortex creates an imbalance that makes the relatively high-pressure air inside the structure burst outward. Since the air can't escape fast enough, the structure virtually explodes. Respecting this aspect of a tornado, Charles Sandford of Xenia, Ohio, broke every window and opened every door in his home when, in April 1974, a tornado approached. After the twister had passed, his was the only house on the block still standing.

There are many conjectures about just how rapidly a tornado can move along the ground. Actually the forward speed of a tornado varies greatly, some moving as slowly as five miles per hour, others ripping across the land at considerably greater speeds.

The average cross-country speed of a tornado is between twenty-five and forty miles per hour. On rare occasions tornadoes have been known to be stationary, but such an event is short-lived. The fastest-traveling tornado on record is the tri-state tornado of 1925. During the last stage of its rampage it reached a forward speed of seventy-three miles per hour!

The awesome lifting power of a tornado can be calculated only by what it can do. In 1937 a man walked into a weather bureau office in the Midwest and asked, "What I want to know is, can the thing happen that I saw happen?" The man had witnessed a tornado lift a railroad engine from one track, turn it around in midair, and set it down on a neighboring track, facing the opposite direction.

Similar events have happened on other occasions. In 1931 a tornado of unusual power struck an area in Minnesota. It performed the Herculean feat of lifting a railroad coach weighing eighty-three tons, complete with passengers, nearly eighty feet into the air before dropping its burden into a ditch. There were many casualties.

Tornadoes have a tendency to skip and bounce as they move along the ground. This skipping motion was responsible for the most remarkable tornado observation ever made by a human being. Will Keller, a Kansas farmer, was working out in his field on the afternoon of June 22, 1928, when he looked up and was startled to see a large tornado heading in his direction. He ran quickly to his cyclone cellar, and just as he was about to close the door he turned back to get one last look at nature's destroyer as it swept down on him. He noticed that the lower end that was sweeping the ground had begun to rise.

Knowing of the tornado's tendency to skip, Keller held his position, ready to drop into the cellar if the tornado were suddenly to dip. As the shaggy end of the funnel passed directly overhead, everything became "as still as death." He noticed a strong gaseous odor, and it became difficult for him to breathe. A screaming, hissing sound seemed to come directly from the end of the funnel. Looking up, he was quite astonished to find himself staring into the very heart of the tornado!

The large circular opening in the center of the funnel was over fifty feet in diameter and appeared to extend straight upward for a distance that Keller estimated to be about one-half mile. The walls were made of rapidly rotating clouds, and the interior was illuminated brilliantly by constant flashes of lightning that zigzagged from side to side. Had it not been for the lightning, he could not have seen the opening or any distance into the interior of the funnel.

He noticed small tornadoes that were constantly forming and breaking away around the lower rim of the vortex. These appeared to be the source of the hissing sound.

The tornado did not hover but quickly moved on. The few seconds of Keller's observations have nevertheless been of tremendous value in providing answers to many of the questions about tornado activity.

From the time of Keller's observation to the present no other person has looked into the funnel of a tornado—and lived!

• •

A Delicate Profession

A surgical operation on the eye is among the most delicate of medical procedures. The surgeon must undergo much specialized training and years after receiving a medical degree may still be in training for this specialty.

The first successful corneal transplant was performed in 1835 by a British army surgeon in India. His pet antelope had only one eye, and its

cornea was badly scarred. The doctor removed the cornea from a freshly killed antelope and transplanted it into his pet's eye. To the amazement of his fellow physicians, the operation was a complete success and his pet was able to see perfectly with its one eye.

Specialization in eye surgery, as demanding as it is, actually goes back quite far in medical history. Cataract operations were performed as early as 1000 B.C. in ancient Babylonia. The fee for both success and failure in such an operation was generally fixed by state law.

As financially rewarding as successful surgery may have been, an eye doctor's profession was extremely hazardous. If the operation was a success, the physician received ten shekels of silver, a large sum for those days. If his hand slipped, however, and he somehow managed to blind the patient, the law required that his operating arm be cut off at the elbow!

If he didn't bleed to death, he usually retired from the practice of medicine.

• •

Dinosaur Nesting Grounds

About 120 million years ago a female Ceratopsian dinosaur scooped out a shallow hole in the desert sands of what is now Mongolia. When the hole was dug, she laid a dozen or so eggs inside it and covered the nest with sand. Satisfied that her task was completed, she turned back to the never-ending business of finding food. Doubtless she rarely wandered far from her nest and periodically returned to fight off egg predators.

Her efforts were all in vain, for the eggs never hatched. A violent desert sandstorm covered the nest with a thick layer of windswept material, closing off the air supply. Eventually the eggs became petrified.

In 1923 an American Museum of Natural History expedition uncovered the nest in the Flaming Cliffs area of Mongolia. Apparently the incident described above had occurred numerous times, for hundreds of egg fragments were found along with many dozens of nests. Some of the eggs were dissected by the scientists and were found to contain the unhatched skeletons of infant dinosaurs.

Over one hundred adult skulls were also found in the area. With such a concentration of a single species, ranging from egg to adult, it is quite evident that the Flaming Cliffs area of Mongolia was at that time the breeding ground for this group of dinosaurs. They probably gathered there by the thousands to mate and breed their young annually. This species has now become one of the most thoroughly studied of all dinosaurs.

Protoceratops, as it was scientifically designated, existed in herds on this prehistoric Mongolian desert. They grew up to six-and-a-half feet long and had an enormous head with a hooked parrotlike beak, which was very likely a defense mechanism.

After laying her eggs, the female probably returned to the herd to feed but always remained in the vicinity of the nest. Parental instincts would prescribe that she periodically return to the nest, and it may be that, like the modern crocodile, she helped her young into the world. It is also very likely that her brood had its share of enemies, as do the young of modern crocodiles.

In one instance the American expedition team of 1923 was in the process of excavating a particular nest when, to their amazement, they uncovered the skeleton of a four-foot toothless dinosaur lying about three inches above the eggs. Later named *Oviraptor* (egg seizer), the animal probably lived by feeding on dinosaur eggs. This particular specimen was in the act of digging up the nest when it was overcome by a violent sandstorm and buried alive on top of the very eggs it had come to steal.

There were other enemies; one known to scientists was a fast-running carnivore, *Velociraptor*. This animal was about the same size as *Protoceratops* but, armed with huge teeth and enormous claws, it probably fed on very young hatchlings rather than on stolen eggs. The carnivore may have been drawn to a nest by either sight or the sound of cracking shells. The same circumstances also attracted the mother *Protoceratops*, and a battle royal would ensue. In all probability the carnivore usually won—but not always.

Aeons ago a mother *Protoceratops* returned to her nest only to find a

carnivorous *Velociraptor* in the act of consuming her young, and a death struggle began. The nest robber managed to kill the mother *Protoceratops*, grasping her bony frill as its own life ebbed away.

In 1971 a joint Polish-Mongolian expedition uncovered their remains. The forelimbs of the carnivore were still holding on to the carnivore's adversary, just as they had in that battle to the death—more than 1,000 millennia ago!

• •

Animal Arms Race

A type of arms race appears to be developing among many species of animal life, particularly the insects. Such a race demands a rapid evolution of improved hunting-defense tactics for all groups of organisms involved.

The monarch butterfly has evolved a slick evolutionary trick that shields it from voracious birds such as the blue jay. In the past monarchs were a memorable morsel for the blue jay but in recent years, probably evolving from the butterfly's diet of milkweed, it sequesters in its body packets of poison capable of sending a naive predator into violent fits of retching. Any blue jay swallowing a monarch butterfly in this day and age is almost immediately taken ill. When recovered, the bird will avoid the monarch like the plague. Of course this doesn't do the original swallowed butterfly any good, but its sacrifice helps preserve the race. And because the blue jay will subsequently avoid any bug that even remotely resembles the butterfly, other insects are protected as well.

The evolving traits of protective body poisons apparently extend to other lowly forms of life. For years it was assumed by scientists that invertebrates such as insects were of too limited intelligence to learn from feeding experiences. Recent experimentation indicates that this is definitely not so, at least for the praying mantis. One research entomologist, May Berenbaum, fed snacks of milkweed bugs to a type of mantis and watched as the predator regurgitated its meal. The mantis very plainly refused a second helping and went so far as to haul away the foul-tasting bug offered as dessert. The conclusion was definite, as the scientist observed: "Obviously they're not so stupid. They will reject noxious prey. They not only perceive differences between edible and inedible bugs, but they learn to generalize."

To test her theory further, Berenbaum and a graduate student raised milkweed bugs on two different diets. One group was fed a diet of milkweed, a plant containing a high concentration of cardenolides, a chemical known to be toxic to many animals. A second group of identical bugs was fed a special diet of sunflower seeds. The results were illuminating.

A mantis encountering milkweed bugs for the first time showed no reluctance in pursuing, capturing, and devouring the hapless insects. Those bugs fed sunflower seeds received the same treatment and were quickly converted into dinner. However, as was expected, the mantises that had eaten milkweed-fed bugs became violently sick and quickly gave up their meal. It was then very noticeable that the sadder but wiser predators learned to avoid and reject even harmless bugs that had been painted by the scientists to resemble milkweed-fed bugs.

The evolution of animal protective techniques has expanded in such a manner that insects can even compete for survival with higher forms of life. An excellent example in the growing competition for survival is one involving the bat.

The bat emits sounds at frequencies of up to 230,000 vibrations per second, well above the human range of hearing. The bat's built-in sonar system permits it to read echoes of these sounds, which bounce off intended victims, and the hunter can thereby judge the location of the prey as well as the speed and direction in which it is moving.

Certain species of moths have developed methods of countering the bat's sonar system. Their own hearing organs are, depending on species, located on various parts of their bodies and are used as a built-in early warning system tuned to the bat's high-pitched cries. The element of surprise is thereby defeated, and the forewarned moth is often able to escape.

Extreme refinement seems to be developing in the tiger moth. This species has developed a way to jam the bat's sonar system by producing a clicking sound that actually mimics the bat's echo-location cry. Thus the information relayed to the bat is false and causes it to swerve off course and miss its prey.

Admittedly research into the learning processes of lower forms of life is only in the infant stage. But what is clearly emerging is that such learning processes may be an important selective force that has permitted some insect species to survive through the ages. More recent research has shown that creatures ranging from scorpions to spiders can also be trained to avoid distasteful prey, much to the benefit of the prey.

• •

"Iceberg Dead Ahead!"

At present about 10 percent of the earth is covered by glacial ice. During the last great Ice Age, which ended 10,000 years ago, over 30 percent of the land surface was covered by ice thousands of feet thick.

With so much ice covering the land it is easy to visualize the northern seas being heavily dotted with enormous icebergs. Considering that the earth was then in a state of refrigeration, the number and size of icebergs were not extraordinary. What was unusual was their durability.

There is now definite scientific evidence indicating that huge icebergs actually reached as far south as Mexico City!

With the earth currently undergoing a warming trend, icebergs are more restricted to the frigid regions. Greenland and Antarctica are still covered by huge, thick masses of glacial ice. As these glaciers move into the sea, enormous chunks of ice are broken (calved) from the main mass and float away into the open sea. Thus are icebergs born.

Because the ice consists of fresh water, which is lighter than sea water, the icebergs float. However, only one-seventh of the entire mass floats above the waterline, the remainder being submerged. The appearance of the

iceberg can be deceptive; what would appear as a moderate-sized chunk of floating ice may in fact be huge and extend much farther in all directions under the surface of the water.

Sophisticated modern instruments and constant patrolling by the United States Coast Guard have reduced the threat of sea hazards from icebergs considerably. Unfortunately such precautions were not available in the past, and the greatest of sea disasters have resulted from collisions of ships with icebergs that frequent the northern latitudes. Who can forget the tragedy of the "unsinkable" *Titanic*?

With bands playing, the world's largest ship sailed for New York on its maiden voyage from Southampton, England, with a load of prominent passengers. Apparently the captain was attempting to complete the voyage in record time, as the giant propellers drove the ship at full power, twenty-three knots.

At 9:00 A.M. on April 14, 1912, the captain received a wireless message from another ship warning of ice fields ahead. The message was put on file, and at 1:42 P.M. he received another warning. Underrating the danger, the captain commanded that his sailing orders remain unchanged. By 11:00 P.M. a number of icebergs had been observed, and still the *Titanic* knifed through the dark sea at full steam.

At 11:40 P.M. the lookout high in the crow's nest saw a huge mass ahead. Immediately pressing the alarm signal, he grabbed the phone and shouted, "Iceberg dead ahead!" Without hesitation the first officer shouted, "Hard astarboard! Engines full astern!" Everyone on the bridge held his breath and stood frozen. There was a slight crunching sound; hardly anyone felt any impact, so they began to feel it had been a very close call. But half an hour later the damage was assessed. The *Titanic* was mortally wounded.

At that time the great ship had a double-bottomed hull divided into sixteen watertight compartments. Since four of these compartments could be flooded without endangering the ship, it was considered unsinkable. However, the iceberg acted like a huge can opener and made a 300-foot gash in the right side, rupturing five compartments, and the ship started to sink.

Distress signals went out over the wireless, and rockets shot skyward. There were many acts of heroism as most of the women and children were placed in lifeboats, and some of cowardice as several of the men jumped into boats being lowered. The *Titanic*'s lifeboats could accommodate only 1,200 of the 2,200 on board, and in the pandemonium many of the boats were lowered into the sea half full. Many passengers jumped into the icy waters, hoping to be picked up by the lifeboats. This was a futile gesture, as the boats were hurriedly rowed away from the *Titanic* to avoid being sucked under. Those splashing in the sea were generally ignored and quickly succumbed to the freezing sea.

By 2:20 A.M. on April 15 the stern of the great ship had risen from the sea and stood high in the air, water cascading from her three huge propellers. The *Titanic* stayed in this position briefly and then slid forward into a watery grave, taking 1,501 passengers and crew with her. At 3:30 A.M. the liner *Carpathia* arrived and picked up about 700 survivors from the lifeboats floating in what was now a sea of the dead.

A sailor on a German ship in the disaster area spotted an iceberg with a long red paint scar running along its base and photographed it. The photo has been preserved, because this is probably a picture of the murderous iceberg.

• •

The Law That Bounced

In the mid-1870s millions of bison still roamed the American West and were a critical food source for wolves, grizzlies, and Indians. The tidal wave of bison slaughter that followed is well known. People reasoned that so vast were the bison herds that they could never be eliminated. A mere decade later only a few hundred bison remained in hidden valleys of Yellowstone National Park.

Entrepreneurs, drawn by the opportunity to cash in on all the free grassland, quickly imported cattle and sheep to replace the buffalo.

It was simple at first, for all the ranchers needed to do was to turn the livestock out into the open range. This success was short-lived, however, because the brutal winter of 1881 wiped out 95 percent of some herds, forcing the owners to retrench their operation. Fences went up, and so did the cost of ranching. The loss of livestock became a critical factor, especially for some cattlemen, who were operating on a low profit margin.

There was little choice involved for the wolf. In finding the range depleted of bison, its natural prey, it turned to another source of food, the plentiful and very vulnerable cattle and sheep. The loss of livestock to the wolf weighed heavily on the livestock operators, and several cattle states quickly placed a bounty on the head of the wolf.

And so the wolf, because of a situation forced on it, became the natural enemy of man, a creature to despise, hunt, and destroy. In Montana alone, between 1883 and 1918, over 80,000 wolves were killed. Hunters shot them on sight, poisoned them, and dynamited their dens. It became very obvious that the time of the western wolf was coming to an end.

By the turn of the century the wolf was plainly the whipping boy that shouldered the blame for the difficult life on the frontier. So intense was the drama that the program of extermination was quickly expanded to bizarre proportions. The Montana State Legislature in 1905 passed a law that required the state veterinarian to inoculate captive wolves with a contagious deadly disease called sarcoptic mange and then release them back into the wilds so they could infect other wolves. This grotesque program was doomed from the start. The infected wolves remained near the source of food. And so the disease quickly spread among the livestock and took a far greater toll than the wolves had ever taken. Montana ranchers had no choice but to hunt down infected wolves and destroy any of their stock that they even suspected of having the disease. A number of cattlemen were forced into bankruptcy.

Despite all this, enthusiasm for wiping out the wolf continued. By the 1930s the wolf was close to extinction in the West, but ranchers felt any wolves were too many, so the federal government sent in its own set of exterminators. Noted hunters and trappers were placed on government payrolls, and the slaughter again began in earnest. These men hunted, clubbed, and trapped the wolves; trapped wolves were quickly shot. Surviving wolves fled to Canada, where the American hunters could not follow. However, the job had been completed, and the wolf was now completely extinct in the Rockies and adjacent territories.

The wolf is now vigorously protected as an endangered species, and a few have drifted back into their ancestral hunting grounds. Once again this is causing unrest among cattlemen. Complaints to the government are beginning to come in, and there is talk of lowering wolf status from endangered to threatened.

Most people just do not recognize the importance of the predator in controlling the herbivore populations. In America the wolf was the chief game warden, as was definitely proven in the Kaibab Plateau of Arizona.

At the beginning of the twentieth century this part of Arizona was home to about 4,000 deer and many wolves. In 1906 the area became a game preserve, so the wolf had to be eliminated. The massacre was so effective that within twenty years not a single wolf was alive on the Kaibab Plateau.

With the wolf gone, the deer population swelled tremendously, and many died of starvation. Unfortunately not enough deer did die, and their population growth was explosive. As their increased numbers caused the environment to deteriorate, starvation became the order of the day. There was only one recourse. Since the wolves, who normally would have controlled the population, were gone, the responsibility of correcting the imbalance of nature fell to man!

• •

The Blue Lake Rhino

Fossil hunters, for the most part, confine their explorations to rocks that were formed as a product of erosion—sedimentary rocks. Igneous rocks, having originated from a molten state, would hardly be expected to contain fossils.

Yet a rather strange incident occurred in an area now known as the state of Washington approximately twenty-five million years ago. At that time the area was vented by prominent fissures many miles long, from which flowed immense quantities of soupy, molten basaltic lava. This material, derived from deep within the earth's crust, spread rapidly, forming huge lakes of burning rock surrounded by mountains. This process of land building continued on and off for thousands of years, creating a basaltic structure now known as the Columbia Plateau.

Animal life during that remote Miocene Epoch was quite abundant, and doubtless many animals fled in all directions during each eruption. Apparently, however, not all the prehistoric denizens managed to escape.

In 1935 a group of workers came across a curious hole at the base of a basalt cliff near Grand Coulee. Poking around inside the hole, they found a few charred bones and fossil teeth of an extinct rhinoceros. A scientific investigator showed the hole actually to be a mold of the ancient victim.

Scientists believe the animal was overtaken by the lava in a small lake that it inhabited. The water caused rapid cooling of the lava, which engulfed the body and preserved its outline in a somewhat crude mold. Quite probably the rhino was dead when it plunged into the lava. However, several scientists believe that in its terror the rhino simply ran headlong into the encroaching lava and was engulfed by the cooling molten mass.

It must have been a source of amusement to the worker who first discovered the "cave," because he poked his head into a twenty-five-million-year-old rhino—through its rump.

• •

CHAPTER THREE

And Then There Were None

Who would have believed that a bird that numbered in the billions could ever become extinct? But when it was found out that it was valuable, the slaughter began in earnest . . .

The passenger pigeon was the most spectacular casualty of unrestrained hunting in nineteenth-century America. In the year 1800 they accounted for about one-third of the world's bird population, yet in slightly over one hundred years their population dwindled to zero.

Early American explorers were greatly impressed by the tremendous hordes of passenger pigeons. During the early 1800s they numbered well into the billions, and their nesting areas extended for miles. Branches broke under their weight, and trees were stripped bare of their leaves; the ground was covered so thickly with droppings that almost every bush and blade of grass was smothered. The birds left a devastated landscape in their wake.

Reliable observers once described an enormous flock of pigeons as they flew overhead. The flock was more than a mile wide and about 250 miles long. They estimated that this single flock contained well over two billion birds. Audubon witnessed a flight of passenger pigeons in 1813 and wrote, "The air was literally filled with pigeons. . . . the noonday light was obscured as if by an eclipse."

Since the passenger pigeon was edible and marketable, the slaughter was big business by midcentury. Audubon once described a hunt, with people firing shotguns into the masses of birds, killing them by the thousands. Since the pigeons were worth $1 each on the open market, professionals were attracted to the hunt. One man boasted of killing 10,000 pigeons per day! With shotguns, traps, nets, and poles, humans managed to extinguish over six billion birds within sixty years—a most remarkable feat!

The growth of civilization completed the job; as cities spread, natural habitats were destroyed. When it became obvious that the pigeon popula-

tion had been decimated, some isolated efforts were made to replenish the flocks. But passenger pigeons, gregarious birds, were disinterested in increasing their ranks when they were paired off. It was almost as if the species was resigned to extinction. By 1900 only a few scattered pigeons remained. Still the shooting continued; and then there were none.

Well, not quite none. On March 24, 1900, near Sargeants, in Pike County, Ohio, a little boy was trying out a new air rifle he had received as a gift. Spotting a large pigeon on a branch close by, he drew a bead and squeezed the trigger. His aim was good, and the bird, shot through the head, died instantly. Everyone was very impressed by the hunting prowess of the boy, and although nobody realized it at the time, he had just shot the last passenger pigeon ever seen in the wild. A taxidermist friend mounted his trophy, but for reasons unknown, instead of returning it to the boy he gave it to Ohio State University where it was identified.

It is truly pathetic that less than a hundred years earlier the sky was blackened by billions of passenger pigeons in flight. But what is most deplorable is that man, with no help from nature, was able to exterminate a species by killing them at the rate of a hundred million per year!

A few of the birds did live on, but only in zoos, a pitiful remnant of a once great species. The last passenger pigeon, twenty-nine-year-old Martha, died at 1:00 P.M. on September 1, 1914, in a Cincinnati zoo. At that precise moment the species became unequivocally extinct.

Dreams Carried by Butterflies

The most popular insect, because of its natural beauty and lifestyle, is the butterfly. Insect collectors seldom bypass one of the typically beautiful specimens. But where superstitions are concerned, butterflies indicate many things to many people, and the interpretations in different localities are often contradictory. To the superstitious of Louisiana the sudden appearance of a white butterfly, especially if it flies around a particular person in that house, means that good luck will follow. In Maryland the same occurrence forebodes death.

The belief that butterflies play a role in weather prediction is quite ancient and widespread. Some people maintain that any butterfly flying in their face is an indication of immediate cold weather. The Zuni Indians of the American Southwest believe that the sudden appearance of butterflies usually foretells of good weather. They would say, "When the white butterfly comes, comes also the summer," and "When the white butterfly flies from the southwest, expect rain."

Probably the most pleasant butterfly superstition was one common among the Blackfoot Indians. They believed that when persons were asleep, dreams were carried to them by white butterflies. Therefore it was customary for mothers of the tribe to embroider the image of a butterfly on a small piece of buckskin and tie it to their babies' hair, thus encouraging the infants to go to sleep. Simultaneously the mother would sing a lullaby to the child, asking the butterfly to come fly about, put the baby to sleep, and give it pleasant dreams.

This belief probably has its foundation in fact. When a person watches a white butterfly for a length of time, its silent, graceful, rhythmic fluttering motion actually does induce sleep. And not surprisingly, the dreams occurring during this somewhat hypnotic sleep are usually pleasant.

• •

Vampires Are Real

The movie and television industry has thoroughly exploited the legend of the vampire, which was made famous by Bram Stoker with the novel *Dracula* in 1897. Strangely enough, vampires do exist.

When vampire bats were first discovered by Spanish explorers in the jungles of Central and South America, they immediately became associated with the medieval legends of the undead, such as vampires and werewolves. The vampire bat of modern fiction is always associated with the undead, and this has become its trademark. The vampire bat of real life is totally

different except for one major feature: it does feed on blood. In fact blood appears to be its sole diet.

These flying mammals are very abundant and relatively small. The largest has a twelve-inch wingspan and a four-inch-long body, but most vampire bats are much smaller. During the day they roost in caves, old mines, hollow trees, crevices in rocks, and old buildings.

Colonies of vampires may include as many as 2,000 bats, but the average colony has close to one hundred. The sexes roost together, and they may even share a cave with bats of other species. Shortly after dark the vampire bats leave their roosts in slow, noiseless flight, usually only one to three feet above the ground because they are seeking a victim who is sound asleep.

When a victim is found, the bat alights alongside, noiselessly crawls up to it, looking like a large, weird spider, and with its sharp incisors makes a quick, shallow bite in a place where there is neither hair nor feathers. Without a sound the bat then laps up the blood. Its victim nearly always sleeps through the entire event. So light is the attack that even a man will sleep through the bat's entire dinner. The common vampire can drink such relatively large quantities of blood that it is barely able to fly for some time afterward.

Despite the bat's gruesome feeding habits, its size guarantees that the actual amount of blood loss sustained by the victim is very slight. The chief danger to the victim lies in infection or the possibility that the bat may be rabid.

Vampire bats seem to exhibit a preference for certain individuals, animal or human, that they have attacked previously. Some scientists believe they actually develop a taste for that particular recipe of blood. They almost seem to follow the habits of Bram Stoker's Dracula in their search for a human victim, because they appear to seek out the same victim night after night. But they depart from the Dracula syndrome in where they bite: they never attack the throat of a sleeping human; instead it's the individual's big toe that they go after. One would think the harassed person would start wearing shoes to bed!

• •

Werewolf!

Even a man who's pure in heart
And says his prayers at night,
May become a wolf when the wolfbane blooms,
And the moon is full and bright

The Wolf Man

Speaking of legends, that of the werewolf is one of the most persistent in the history of civilization.

The belief that a man can become a wolf under certain conditions is ancient and very strong. Throughout history the contemporary police have actually attributed many recorded murders to werewolves.

Even today there are, in a few isolated sections of Europe, people who still fear the light of the full moon and refuse to venture into the darkness of night. Their reasoning is quite explicit: this is the time when the lost soul walks in the form of a wolf searching for a human victim.

Surprisingly there does exist a certain disorder of the mind that strongly resembles this strange superstition, and it may have been the cause of the original belief in werewolves. The disease is known as *lycanthropy.*

Medically speaking, lycanthropy is a form of acute mental illness in which the victim actually believes he is a wolf. While in this demented state, he will imitate the physical activities of a wolf on the prowl and can be dangerous. Rare as this disease may be, there are several cases being treated by psychiatrists at the present time!

It can be said, however, with little fear of contradiction, that the only place that a man can be transformed physically from a man to a wolf is in Hollywood.

• •

Ordeal!

Just a few years ago three geologists were encamped on the bottom of the Grand Canyon. They had hiked a considerable distance in studying the varied rock formations and retired early, each cozy in his own sleeping bag. Later two of the men were rudely awakened by the third, who was screaming in terror. He appeared to erupt from his sleeping bag and, picking up a large piece of firewood, pummeled his bed vigorously with it. A chilled snake had crawled into his warm sleeping bag for the night. Fortunately it was a nonpoisonous specimen, and the scientist suffered only from fright. But not all naturalists have gotten off so easily. The following is an account of a more harrowing experience.

In 1949 a group of American scientists set up camp in the jungle hill country of the Canal Zone. They slept in sleeping bags under canvas rain shields. At about midnight one of the men heard a faint rustle near his head, and moments later a large snake slithered across his cheek. It crawled into his sleeping bag, and when it reached his stomach it coiled to spend the night. The man froze instantly with the sound of the first rustle; he suspected the snake was the deadly bushmaster common in the region.

Through the long, dark hours of the night the man remained absolutely motionless, knowing that any movement might mean an agonizing death.

At dawn his companions, who at first chided him for oversleeping, drew up abruptly when they saw his gaunt face and the suspicious bulge in his sleeping bag. One of the Indian guides gently cut open a slit at the foot of the bag and blew cigarette smoke into it. The smoke drifted out of the top of the bag, bringing tears to the eyes of the trapped man as he struggled not to cough or move. This tactic was less than effective, but no other way to motionlessly encourage the snake to wake up and leave occurred to the crew.

The snake had now been in the sleeping bag for about twelve hours, during which time the man had not dared to move a muscle. The day warmed rapidly, and the man began to sweat in the bag; finally the snake stirred, but settled again.

One of the trapped man's companions carefully removed the overhanging rain shield, allowing the tropical sun to beat directly onto the sleeping bag. By now the man was nearly delirious from the merciless heat, and almost every muscle was cramped; still he did not move.

Just when he was on the verge of losing consciousness, the bulge in his stomach region moved. The heat had finally become too much for the visitor, so it crawled out of the sleeping bag and headed for the shade. It was promptly killed and held up for all to see—a large, deadly bushmaster.

Ironically, now that he could move again, the man passed out. He was revived and carried to the hospital for treatment. For him the ordeal was over, but as soon as he recovered he resigned his job and left the Central American jungle. Very determined never to repeat such a night of terror, he relocated to a desert environment.

• •

The Man Who Died of a Toothache

Because many of today's dental problems are the result of refined foods and sugar, many people believe that prehistoric man, who ate only natural foods and no refined sugar, was free of dental decay. Unfortunately for early man, this was not always true. Here follows an unusual account.

The limestone mass known as Broken Hill in Northern Rhodesia, now known as Zambia, stands about fifty feet above the surrounding plain and contains numerous caves. One in particular slopes downward for about ninety feet and, when discovered, was almost completely filled with fossil animal bones and cave debris. It was also highly impregnated with ore minerals of lead, zinc, and vanadium and was quickly exploited for its mineral wealth.

In 1921 two miners cutting into relatively soft ore with picks were astonished as they suddenly exposed the skull of a man staring at them through eyeless sockets. Fortunately the skull was preserved in the mine office for scientists to examine. Rhodesian man, as he was later called, had not been buried ceremoniously in the cave upon his death about 100,000 years ago; he had either been thrown there as refuse or carried in by an animal.

Of considerable interest was the condition of his teeth. Ten of the fifteen present were in varied stages of dental decay, including several with abscessed roots. Studies of the pathological features of the skull indicate that the septic condition of the mouth had spread sufficiently into the skull to have actually killed him!

Some of the human bones found near the skull showed stages of rheumatic growths. Scientists believe that, if they belonged to the owner of the skull, this disease may also have been related to his dental problems.

Rhodesian man was a primitive type of Neanderthal. In life he must have been a rather powerful man, standing about five feet, ten inches tall and weighing about two hundred pounds. This particular individual must have had a frightful disposition and temper, but who could blame him? He had a perennial toothache!

• •

Joey

The pouched marsupials of Australia and of Central and South America give birth to young that are more like embryos than true infants. These

tiny immature offspring, although blind and naked, are born with the life-preserving instinct of getting from the birth opening to the mother's pouch.

A marsupial's pouch is like a living incubator for the infant; here it will rest and feed by attaching to one of the teats until it's mature enough to explore the outside world.

Marsupials are probably the forerunners of placental mammals, and they were quite widespread throughout the world in the geologic past. With the development of the placental mammal competition for survival was too great, and marsupials became depleted in Europe, Asia, and North America. They survived in Australia and South America because, as placental mammals steadily evolved throughout the rest of the world, these land masses became islands isolated by marine barriers. Then, as the American continents became connected via a land bridge, placental mammals invaded South America. Only the tough, generic, no-frills opossum survived the placental invasion and even joined the modern mammals in North America.

The mode of life of marsupials is very similar to that of the placental mammals and clearly illustrates parallel development. There are marsupials that are catlike, doglike, mouselike, and eaters of flesh, plants, or insects, as well as some that graze, hunt, burrow, climb trees, and even glide.

Probably the best known of the marsupials is the kangaroo, whose breeding habits have been studied thoroughly. In Australia the young of the kangaroo is known as a joey.

About a month after mating, the female gives birth to her young. The newborn kangaroo is almost transparent, weighs about 0.03 ounce, is 0.8 inch long, and could easily fit into a teaspoon. Although it is blind, hairless, and poorly developed, its front limbs are quite strong. With these it scrambles and claws instinctively up a path in the fur inadvertently licked by the mother as she cleans up after release of fluids from the ruptured egg membrane. In about three minutes the infant climbs into the pouch, where it attaches to a teat; here it will remain attached for the next few months.

Six to eight months later the young kangaroo pokes its head outside of its living cradle and observes the outside world for the first time. Soon it will be nibbling grass along with the mother, but will leap headfirst back into the pouch at the first sign of danger. Even after it is a year old, the young joey will still attempt to hide in the mother's pouch. By now, however, it is quite a tight fit and rather uncomfortable for Mom. When the mother kangaroo has had about enough, she grabs hold of the youngster's tail and flings her offspring away. The joey will make a number of attempts to reenter its former home, but the results will always be the same. Eventually the young kangaroo will take the hint—joey is now on its own!

• •

A Noble Youth

In the year 76 B.C. a group of Mediterranean pirates captured a youth who gave all the appearance of being of the nobility. Convinced that they had captured a rich prize, the pirates set his ransom at twenty talents (equivalent to about $10,000 today). The boy was actually insulted by their offer and laughed in their faces, insisting he was worth much more—at least fifty talents. Brave to the point of foolhardiness, the youth threatened that if his ransom was met he would return and have them all hanged.

The ransom was paid, and true to his word, the boy guided a naval expedition to his former captors. Pocketing the fifty talents, he calmly watched while each of the pirates was hanged for his crime. The boy's name? Julius Caesar.

• •

Portuguese Man-of-War

In 1941, on a sweltering August afternoon a boy ran screaming from the surf off Rockaway Beach, New York. Bathers tore at the bluish strings that were draped over the boy's neck and shoulders, and soon they too were writhing in agony. The boy had encountered a Portuguese man-of-war.

Viewed from shipboard, a fleet of *Physalia* makes an impressive sight. They look like iridescent blue-purple balloons floating gracefully on a white-crested sea. To the British sailors of three centuries ago they looked like miniature Portuguese galleons and thus were named the Portuguese man-of-war.

The float is really part of an enormous colony of jellyfish all living together as a single organism. Each member of the colony performs a particular function ranging from reproduction to searching for food. Just under the float are digestive polyps whose sole function is feeding. The digestive nutrients are distributed to each member of the colony.

Hanging from the float are many hunting tentacles that may range up to sixty feet in length. These tentacles bear thousands of stinging capsules that secrete a poison as potent and toxic as that of a cobra. There is no known antidote to it, and a human, depending on health and age, can succumb in minutes. Even a beached *Physalia* is dangerous, and to step on a dried blue string will cause an excruciating hotfoot or worse.

Like the rhinoceros groomed of insect pests by the tickbird, the *Physalia* has its own symbiotic relationship. *Nomeus*, a small fish immune to the deadly poison, darts in and out of the mass of tentacles. Swimming in circles, it lures larger fish looking for a meal into what must appear to be a forest of seaweed. Here the victim is quickly overwhelmed by the lethal

tentacles. Thus little *Nomeus* procures prey for the jellyfish and in return is protected and fed by tearing pieces from the victims.

The immense jellyfish is well adapted to survive harsh weather without any particular strain. During a storm *Physalia* deflates its bladder, thereby losing buoyancy. It then sinks below the surface, safe from churning waves until the sea is calm again. Within a few minutes it can reinflate itself by producing gas from a special gland.

As deadly and invulnerable as this jellyfish appears, it does have a natural enemy—the huge, lumbering loggerhead turtle. Five-hundred-pound turtles have been observed plowing through colonies of *Physalia* with their mouths trailing strings of tentacles like blue streamers from a maypole. In their wake the water is clear, but on either side are bobbing blue floats and streamers that are at last quite defenseless. The Portuguese man-of-war has met its match and lost.

The Reluctant Giant

Charlie Byrne, a twenty-one-year-old Irishman, stood eight feet, two inches tall in his stocking feet. Byrne was accustomed to people staring at and following him; in fact he made his living by charging a fee for a look at his enormous form. He billed himself in the sideshows of London as "The Irish Giant, a Living Colossus." He was nevertheless a little apprehensive when he noticed a particular gentleman lurking behind him.

Why should this stalker have bothered him? Only because Byrne recognized the man as a professional body snatcher, currently in the employ of one John Hunter, an eminent Scottish surgeon and anatomist. In truth Dr. Hunter did want Byrne's bones and had sent his ghoul around to make the giant an offer; cash now for his body later. Horrified at the idea, the giant quickly hired an undertaker to make him a lead coffin and arranged upon his death to be buried at sea. In this manner the bones of Charlie Byrne would not be desecrated.

The giant knew that careful planning was prudent if he was to avoid becoming a permanent specimen, because death was not far off. Charlie suffered from tuberculosis, an affliction compounded by acute alcoholism. And so in 1783, after a severe drinking binge, the giant developed pneumonia and passed on to his reward. He died presumably tranquil in the knowledge that he had kept Dr. Hunter at bay.

Things did not, however, work out in accordance with his plans—not in any way. While Byrne lay dying, Hunter was not far off, busily bribing the undertaker's employees. And so after Charlie's death the anatomist obtained his body and proceeded to extract the bones. Unfortunately the doctor seemed to regard his remains more as a curiosity than as a source of anatomical enlightenment. It wasn't long afterward that the bones of the former giant were transported to London's Royal College of Surgeons Museum, where they are on display to this day.

In 1909 the American neurosurgeon Harvey Cushing visited the museum and was intrigued by the bones. His examinations of the remains of the reluctant giant revealed substantial information. Cushing opened Byrne's skull and found evidence that during his life there had been a tumor on his pituitary, a tiny gland located at the base of the brain. Scientists now know that the pituitary gland secretes a human growth hormone. If this hormone is released in overabundance, giantism will result, and this was the cause of Charlie Byrne's tremendous size.

Dr. Cushing's observation helped link the pituitary gland to human growth. Thus was launched the modern science of endocrinology.

• •

The Mountains That Weren't There

Mirages are illusions created when light passes through adjacent layers of air of different temperatures and densities. The light is bent, creating deceptive images. Many people believe such illusions occur only in hot, dry deserts. This is certainly far from true, for mirages have been seen even in the Arctic.

During the early part of the twentieth century the American polar explorer Robert Peary reported the existence of an unmapped Arctic mountain range. "We saw the mountains and called them Crocker Land," he announced.

Through the years that followed, the mountains aroused worldwide curiosity. What treasures did these mountains contain? Did unknown tribes of people live there? In 1913, the American Museum of Natural History mounted an expedition in search of Peary's Crocker Land.

Far into the icy sea of the Arctic the expedition ship sailed. But where Peary said the mountains were located the explorers found only icy wasteland. Undoubtedly a miscalculation, they surmised, and the search continued. Eventually the explorers did find Crocker Land, but to their astonishment it was nearly 200 miles west of where it had been presumed to be.

Sailing as close as possible, they dropped anchor, and a select crew of men set out on foot over the ice. As they approached the mountain range, it receded. When the men stood still, so did the mountains, always beckoning to them. In the polar sunlight they could see dark valleys promising great mineral wealth.

The men redoubled their efforts and approached a valley enclosed by mountains. They felt sure of success, but by now the Arctic sun had begun to sink below the horizon. The men stood dumbfounded watching the mountains dissolve with the setting sun as if by magic. They were standing on a vast, flat expanse of ice. As far as the eye could see, there was nothing but ice in all directions—Peary's Crocker Land was a mirage!

Nature's trickery with mirages hasn't always been so inhumane. The illusion created by one particular mirage resulted in a most merciful incident during World War I. One day a British officer ordered an artillery barrage on a Turkish position. Before the order could be carried out, an illusory landscape suddenly appeared before the eyes of the artillery men. It became necessary for the officer to rescind the order; the mirage had completely masked the enemy's location.

The illusions created by mirages include an amazing array of distorted views, from the imaginary mountains of "Crocker Land," to the "refreshing" puddles of no water on desert highways, to the delightfully entertaining vision of two Eiffel Towers, one balanced neatly on top of the other, upside down!

• •

Zeus Is Laughing

During a great thunderstorm over England in July 1923 over 6,000 lightning flashes were recorded in London alone, which included 47 in one minute. Perhaps 1923 was a special year, because on Christmas Day of that year a storm over Pretoria, South Africa, produced about a hundred flashes per minute for more than an hour.

Such frequencies of lightning should not be surprising. In the few seconds it takes the reader to peruse this paragraph, lightning will have struck the earth about 700 times. This rough estimate of one hundred times per second of discharge on a worldwide basis represents over four billion kilowatts of continuous power!

Lightning is one of the most powerful forces of nature, causing almost incalculable damage—and, too often, death. In the midst of death and destruction this heavenly electricity often produces unusual and even comical events. If the ancient historians were to review some of these phenomena, they would probably say, "Zeus is laughing."

This would be understandable, because the ancient Greeks believed that bolts of lightning were hurled by the god Zeus, and they actually worshiped the ground where it struck. They often built fences around the burned area and looked on it as a religious shrine. Guards were posted around large strike areas to make sure the passersby would worship the shrine properly. Disrespect for Zeus was dealt with harshly.

The divination of lightning did not stop with the ancient Greeks, but became part of the Roman religion and continued into the Middle Ages. As late as the eighteenth century a thunderstorm was the signal for a furious outbreak of bell ringing that was supposed to have a soothing effect on the gods of the storm. Medieval bells were frequently inscribed FULGURA FRANGO—meaning "I break the lightning." Bell ringing became an honored, but dangerous, occupation. When a storm approached, the honoree had to ascend the church tower and furiously ring the tower bells. A number of men gave their all for the cause; many times lightning would strike the church tower, and the bell ringer would be the first casualty.

Some of the freak events resulting from bolts of lightning seem to suggest strongly that Zeus is at play and having a ball. Some years ago in Idaho several lightning bolts struck a field of potatoes and burned the stalks to cinders. But the potatoes underneath were cooked to a turn, just as if they had been cooked beneath hot ashes! They retained their heat for a while and were soon served up as baked potatoes.

This must have pleased Zeus, because in 1943, a soldier on maneuvers retired for the night into his sleeping bag. A sudden bolt of lightning struck the zipper of the bag and completely welded the soldier inside. The man was frightened and scorched but otherwise unharmed. When cut from the bag, he is said to have remarked, "Now I know what a baked potato feels like."

In another incident lightning set fire to a building, but the same bolt bounded off and struck a nearby fire alarm, thus calling out the fire brigade to fight the very blaze it had started. This was a lightning bolt with a conscience. In the same storm lightning struck a chain maker's shop, causing several chains to become solid bars of iron.

And then there was the time Zeus decided to play golf. On August 10, 1977, a sixteen-year-old boy playing on a Florida golf course was caught in the open during a violent thunderstorm. While he ran for cover, a bolt of

lightning struck the golf umbrella he was holding. The umbrella was shredded along with the boy's shoes, and he was rendered unconscious. The doctors who attended him stated that his pulse rate had increased to 200 beats per minute! However, Zeus was only playing and the boy recovered.

Zeus did not neglect the shoemaker. In France some years ago lightning struck a cobbler's shop. This affected the craftsman's tools in such a manner that his hammer, pincers, knife, nails, and other metal implements became magnetized and were constantly sticking together. So unnerved was the shoemaker that, being a religious man, he brought the local priest to his shop and demanded it be exorcised.

In many instances lightning has been known to magnetize objects so powerfully that they are capable of holding three times their own weight.

Recently a number of stories of Zeus at play have come to light—many are probably somewhat exaggerated or distorted to provide a story with a message. There is the case of lightning striking a British warship at sea and melting the gold braid off the unpopular mate's uniform. In Argentina there is a record of lightning striking a man's bedsprings at night, so deftly that the man sagged to the floor without waking up. Perhaps the crowning story of Zeus at play concerns a brewery that was hit in such a manner that the beer was instantly aged, flavored, and ready to market. In a recent storm lightning caused a large electric clock to stop. Minutes later it started again, but the hands went backward at twice the normal speed! Lightning once struck a room where a girl was sitting at her sewing machine. In a brilliant flash of light the girl found herself sitting on top of the sewing machine, completely bewildered but unharmed.

At times Zeus seems to be toying with people. Men struck by lightning have sometimes found themselves unharmed but completely disrobed, their clothes scattered in fragments over a wide area. Recently two women were walking down the street during what seemed to be a mild storm in a midwestern American town. Their clothing was suddenly torn away by a lightning bolt, leaving them disrobed but clearly unharmed; their bodies were not even scratched. Zeus had foreseen this, and instead of leaving them completely naked, he left them both still wearing their shoes!

Despite these occasional tricks by the god of thunderbolts, a person struck by lightning is usually killed instantly; but some do survive and even beat the odds by being struck twice. A most phenomenal case on record concerns a Shenandoah National Park ranger who has actually been struck by lightning seven times. He is famous in Virginia and has naturally been dubbed "the human lightning rod."

The man bears scars on his right arm and leg from the first strike, which happened in April 1942 while he was fleeing a fire tower during a violent thunderstorm. When the man was struck the second time, in July 1969, his eyebrows were burned off. His left shoulder was badly burned in July 1970 during the third strike. On the fourth occasion, in April 1972, the man's hair was set afire. The fifth strike was more violent; the man was knocked more than ten feet out of his car, and his hair was again set on fire. He was injured but survived his sixth strike in June 1976. In June 1977, while he was fishing, lightning struck the man for a seventh time and put him in the hospital with severe burns on his chest and stomach.

During a hospital interview he stated, "You can tell it's going to strike, but it's too late. You can smell sulfur in the air, and then your hair will stand up on end, and then it's going to get you."

Scientists cannot even come close to explaining why this phenomenon can happen to a single person so many times. Perhaps there is something in the park ranger's body chemistry that causes it to act as a lightning rod. What is even more phenomenal is that the man survived each strike. Zeus, in the middle of nowhere, must be snickering and remembering an adage from the ancients: "He whom the gods would destroy they first make mad."

This is probably hitting close to reality, for the man from Virginia must be at least apprehensive when caught in the open during a violent thunderstorm.

• •

The Eagle and the Hawk

During the early stages of the First World War, French pilots often took bricks aloft on their observation flights. When passing a German plane (which was usually on a similar mission), they would hurl bricks at the enemy aircraft. The objective was to hit the opponent's propeller with the brick. Not surprisingly, the German pilots frequently retaliated.

Several of these early observation planes were actually brought down by the thrown bricks, but there is no record of the damage done to structures on the ground by bricks that missed their targets.

For some reason the brick dogfights led the French high command to decide to train eagles to attack the slow-moving German aircraft during the early stages of the war. The idea was to get the eagle to rush at the plane, which, being somewhat fragile, could easily be brought down along with the sacrificial bird. An entire series of plans was drawn up, but for practical reasons they were later abandoned.

Implausible as it may seem, the fact is that eagles have always been prone to attack slow-moving aircraft, and several cases in which the attacking eagle was successful have been recorded. In this connection it should be remembered that an eagle can weigh up to fifteen pounds and can dive at a speed approaching 200 miles per hour.

A three-motored plane in the late 1940s was attacked by two eagles simultaneously. One flew straight into the middle engine while the other dived from 10,000 feet and went through the metal wing like a rock. A great hole was ripped in the wing, and the aircraft crashed.

Perhaps the most remarkable incident of birds versus machine was a dogfight between an eagle and a Fiat fighter plane during the Italo-Abyssinian War. Screaming as if to announce the attack, an eagle dived on the plane, forcing it to take evasive maneuvers. The eagle and the "hawk" flew around and around, jockeying for position; occasionally the pilot got off a burst of machine gun fire, but never on target. The bird made a sudden dive straight at the plane, smashed the windshield, and struck the pilot on the head. Both went into a tailspin and crashed. The incident is well recorded, because the pilot managed to survive the most unusual dogfight in history!

• •

The Unicorn Was a Whale

In July 1577, Sir Martin Frobisher's ship, seeking shelter from a storm, limped into an inlet at the southeastern corner of Baffin Island. Here the crew found a large dead "fish" with a horn two yards long growing out of

its snout. They believed that they had found a marine species of the unicorn—the counterpart of the terrestrial horned horse.

Few animal products have ever inspired man's imagination as did this whale tusk. Centuries before the time of Martin Frobisher the tusks of this sea creature were introduced on the European continents by Vikings. They knew the origin of the tusk but kept it a dark secret. The impact was quite dramatic, for around these unusual hornlike teeth developed the many myths of the unicorn—a horselike creature with a single horn mounted on its head.

Despite the fact that nobody ever saw a live unicorn, few doubted its existence. During the Renaissance the special properties and magical powers attributed to it were wildly extravagant. People believed that the horn, as well as being an aphrodisiac, had the power to heal, prevent disease, and, above all, counter the effects of poison. To the nobility, who often feared murder by poisoning, the ivory tusk was worth its weight in gold. Four centuries ago Charles V of England gave two unicorn horns as payment for a debt equal, by modern standards, to $1 million.

The myth of the unicorn was fast to evolve and took a long time to die. It survived well into the time of the Renaissance, and scholars finally exposed its true origin in the mid-1600s. Even with this exposé the trade in the mythical horn and the legend of the unicorn continued for at least another hundred years.

Today scientists know that the unicorn horn is a tooth that grows only on the male narwhals. It is the upper left canine, the only tooth that is not lost before maturity. It grows continuously throughout the life of the whale and protrudes through the upper lip as a twisted ivory shaft up to eight feet long.

• •

Minerals for Fools

One of the first minerals that students of earth science learn to recognize is the iron and sulfur mineral known as pyrite (FeS_2). A rather common mineral, it occurs all over the world and is often referred to as "fool's gold" because superficially it resembles the precious metal.

The origin of its nickname has been traced to the California gold rush days. Most of the forty-niners, who included many wanderers, criminals, and misfits, had very little knowledge of the occurrence and appearance of the precious metal in the field. Therefore it was not unusual for a prospector to come upon a ledge of pyrite and think he had struck it rich. These inexperienced unfortunates would rush into town to file claims and buy drinks for everybody. When the assays were returned, the fool-with-the-gold would be informed that his ore was common iron pyrite, worth about $1 a ton. Many townspeople took advantage of the free drinks even though they knew the pyrite was gold only for fools.

Another metal with a somewhat similar nickname was discovered in Colombia almost 500 years ago. It was considered quite worthless and people looked on the metal as "unripe gold," referring to the mineral as "fool's silver." With this attitude firmly established, the miners simply threw it onto the nearest scrap heap, where they expected it to ripen into gold. Some of these "scrap heaps" grew to considerable size.

Today nobody makes unkind remarks about platinum. In recent years, the mining town of Quibdo in the Chocó region has virtually been leveled by treasure seekers searching among the foundations of buildings for nuggets of "unripe gold" that had been so casually tossed aside by early prospectors. The metal has currently become the object of speculative frenzy, and the price of platinum at present is higher than that of gold.

Yet another episode in the tale of fools and their minerals took place in 1532, during Pizarro's expedition to Peru. Some of the explorer's soldiers, always on the lookout for treasure, obtained a cache of emeralds from a group of unfortunate Indians who were promptly slaughtered. These gemstones were reputed to be as large as pigeon eggs. Apparently the conquerors were unfamiliar with this particular mineral, believing emeralds could not be broken. This was a most disastrous bit of misinformation, because when the soldiers pounded them with hammers, the minerals naturally shattered. The men discarded the remainder of the cache, fully confident that they were tossing out worthless colored green glass!

• •

A Case of Antique Murder

However primitive humans may have been two million years ago, they were already showing traits consistent with modern civilized lifestyles. Murder and warfare appear to have been among them.

Quite recently the fossil jawbone of a twelve-year-old *Australopithecine* male was unearthed near Makapan, South Africa. An examination of the jawbone yielded definite evidence that he had died a violent death. His jaw was broken on both sides, and the front teeth were missing. There was a dark, smooth dent on his chin from a violent blow. A clublike weapon, probably the limb bone of a deer or an antelope, had been used. Since there was no sign that the bone had healed, the boy must have died from the blow.

It seems unlikely that an immature boy would have been engaged in active warfare. More probably he had either snatched a morsel of food from an adult or wandered into another group of Paleolithic men. Whatever may have provoked the violence, the boy was killed as a result of it. Since this happened nearly two million years ago, some scientists refer to the incident as a case of antique murder.

Recent studies of *Australopithecus* skulls clearly indicate that the humans of millions of years ago engaged in open warfare against their fellows. A number of skulls examined from several sites in South Africa show that death resulted from skull fractures caused by heavy blows from clubs often fashioned from animal limb bones or, in some cases, from rocks. In one instance a rock two inches in diameter was found inside the skull. Evidence clearly indicates the man had been killed by this rock, which must have been hurled by a powerful hand.

Such open warfare was probably not uncommon, since ancient man appears to have been quite family-clan oriented. Nuclear family distinctions within the clan were doubtless rather vague, for all too often dependent children whose parents had died (for the life span was less than twenty years) would be the responsibility of others in the group. Estimates of the number of *Australopithecines* in a band range from a dozen to fifty individuals. The group would necessarily be large enough to provide protection, and small enough that procuring food and water for all members would be feasible. Clans made up of close family members would lead to territorial hunting and water claims, and instant war was an almost inevitable result when one group moved into another's territory.

Doubtless *Australopithecus* was omnivorous, but hunting was definitely part of his lifestyle. In South Africa two million years ago his favorite prey appears to have been baboons. Scientists have recovered fifty-eight baboon skulls from a cave in South Africa. All had been clubbed to death.

Of the fifty-eight skulls only eight were hit from behind. This must have occurred as they turned to flee from the savage apeman wielding a club, a weapon that had once been the limb bone of an antelope. The fifty others did not die so easily; all the club marks are frontal. Although they probably died at different hunts, all turned and faced their enemy. What a scene of savage fury this must have been! Quite probably not all of the human creatures involved in the confrontation lived to walk away from the carnage.

Because these early humans were small in stature, less than five feet tall, and their average weight was about ninety pounds, one would tend to consider them delicate. On the contrary, they possessed enormous strength, as one hyena discovered abut two million years ago. On that occasion a man responded to the sudden attack by a hyena by thrusting an antelope leg bone into its open mouth. Such tremendous force was exerted that the skull at the back of the throat was broken, instantly killing the beast. The hyena's skull is now exhibited in a South African museum, with the leg bone still protruding from its mouth.

Research on the skeletal remains of a number of *Australopithecine* people has shown that, in addition to possessing enormous strength, they were extremely fleet of foot, probably far more so than today's Olympic runners. An analysis of *Australopithecine* bones indicates that most likely the hominid was able to outrun many of his four-footed contemporaries. He may even have actually run his prey down just as the cheetah does today.

Why have humans so significantly changed in physical potential? Opinions vary, but many scientists believe that it is time rather than mere genetics that has slowed and weakened modern man. The limited demands made on humans physically have reduced the need for such great strength and speed. All agree, however, that an ever-enlarging brain has kept man equal to the task of manipulating and expanding his environment.

• •

Leopards Prefer Dogs

A favorite prey of both African and Asian leopards is the ordinary domestic dog, and these cats seem to go out of their way to capture Fido. One evening in a farmer's home in Kenya a man and wife were sitting in their living room, the man reading a newspaper and the woman knitting.

To complete the domestic scene their dog lay sound asleep on the floor between them. The farmer was just starting to turn the page when a leopard leapt through the open window, grabbed the sleeping dog in its jaws, and was out the window in a flash. All this occurred before the man completed the act of turning the newspaper page. The man recalled only a spotted blur. Needless to say, the couple never saw their dog again.

On another nearby farm a farmer noticed a leopard lying near the edge of his property while several large puppies played close by. The giant cat appeared to be completely ignoring the dogs. The dogs eventually became curious and slowly but cautiously approached the leopard to investigate. The leopard continued to ignore them, making them even more curious. The farmer himself was simply fascinated by what he was observing, because the leopard even rolled over on its back and seemed to be purring. This was too much for one of the puppies, which walked over and sniffed the leopard. That was its last act on earth. The carnivore's reaction was so fast that, again, like a spotted blur, it swept the hapless puppy into its jaws and sped away with its prey. Such clever deception as displayed by this leopard had thus far been unheard of. For a long time afterward the farmer went about his business armed with a high-powered rifle. This did not stop leopard attacks on his dogs, which were, as often as not, successful.

A grand finale to the tales of leopard luncheons has been told by a plantation owner in Malaysia. He was sitting in the tub in his bathhouse, with his pet dog lying on the floor next to the tub. For some reason or other the man had left open the door that faced the nearby jungle. While soaping himself down he suddenly became aware that there was some other creature in the room; it was a large male leopard. In the jaws of the predator was his dog, which was still struggling. The beast had entered so quietly that it didn't become noticeable until it grabbed the man's dog.

The man yelled and splashed water on the beast. The leopard just glared at the man, then slowly turned and walked out of the door and disappeared into the jungle with the dog still in his jaws. The man never saw his pet again. Later, when relating the incident to his friends, he remarked that he couldn't help feeling that the leopard knew he had caught the man with his pants down.

• •

Don't Start Packing Yet!

Oracles and prophets of doom routinely predict that a major earthquake in the western United States will cause California to slide into the sea. The highlands of Colorado will then become the new shoreline overlooking the Pacific Ocean.

In 1969 the soothsayers of California made a serious and most convincing prediction of that nature that caused much unrest among the gullible. In fact a number of people packed up and hastily moved to Colorado. The prediction was that in April of that year a devastating earthquake would

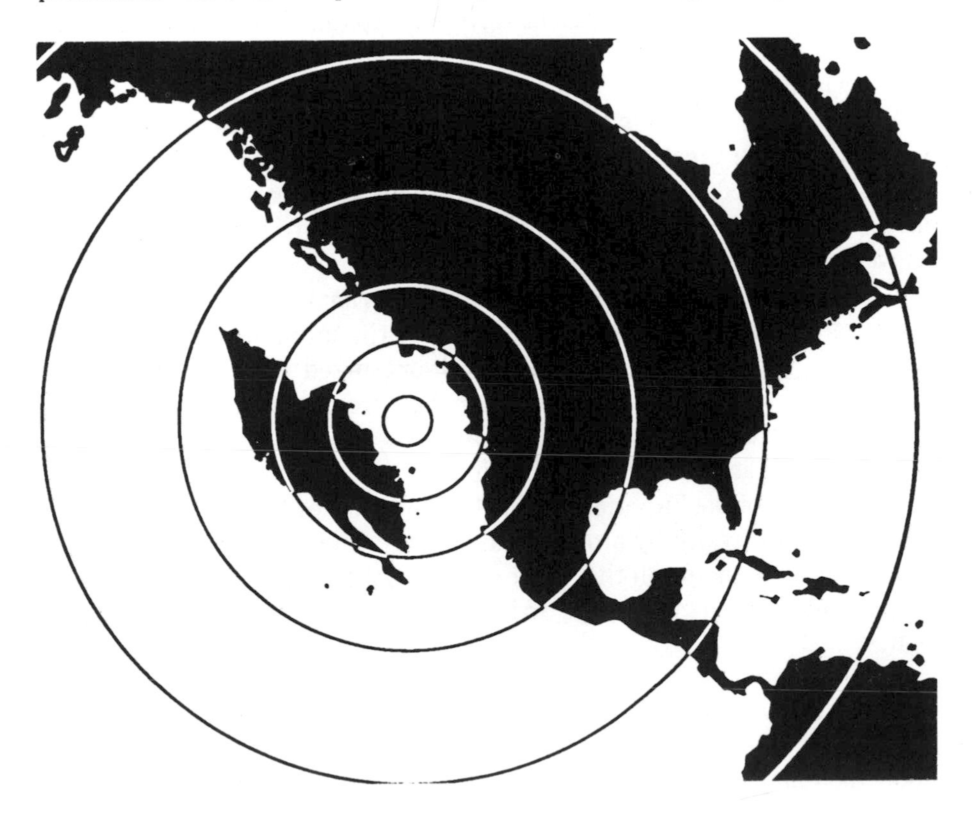

occur; California, in its death throes, would slip into the ocean.

The notion of California sliding into the sea is really quite absurd, if only because of the unshakable fact that the ocean off California is less than two miles deep, while California is over twenty miles thick. Really, it just wouldn't fit.

No earthquake shook the land that April anyway, but the doomsday prophets came up with explanations that seemed quite believable to those gullible enough to have listened to the seers in the first place. Those easily deceived are often satisfied with "The spirit changed its mind," or "The calculations have been adjusted; it will happen next year," or decade, or century. As a consolation prize for those who still insist that California will eventually be broken off into the Pacific, geologists do predict that the land surface now known as California could erode away—in about fifty million years.

• •

Was the Wealth Shared?

The Valley of the Kings is a rocky wasteland on the western bank of the Nile River across from the ancient Egyptian capital of Thebes. Here it was that most of the great kings and nobles were buried in the utmost secrecy to prevent grave robbing.

Belief in the hereafter was strong among the ancients, especially in regard to royalty. Great pains were taken to provide unbelievable riches and to anticipate needs in the life after death. This was clearly reflected in the incredible expenditures that went into the building of tombs and temples as well as their fabulous contents.

It was the Egyptians themselves, motivated primarily by greed, who were the first to desecrate their dead and destroy the mummies of many great pharaohs. Tomb robbery became a well-organized business in ancient Thebes. The tombs were ransacked for their treasures by cunning, well-armed grave robbers, often working in close collaboration with corrupt, greedy priests and well-bribed officials.

By the end of the Twentieth Dynasty almost all of the royal tombs had been opened illegally by professional thieves. The depredations by tomb robbers were so thorough that most of the royal treasures vanished forever long before early archaeologists arrived to complete the process of destruction.

It seems ironic that, despite the great pains taken for secrecy in site location, every tomb had been ransacked and stripped, that is, all but one.

In 1922, when Howard Carter peered through a slit he had made in a door, his hands quivered and his breath was short. He knew he was about

to uncover lost history. He thrust a lighted candle into the slit and gazed in amazement. Standing behind him, Lord Carnarvon anxiously inquired as to what he saw. Carter replied, "I see wonderful things." He later recorded, "As my eyes became accustomed to the light, details of the room within emerged slowly from the mist, strange animal statues and gold—everywhere the glint of gold." He had discovered the almost undisturbed and long forgotten tomb of Tutankhamen.

Tutankhamen, the boy king of ancient Egypt from 1348 to 1339 B.C., was the only pharaoh whose tomb was discovered intact. Ancient grave robbers had actually entered the tomb but never reached the burial chamber.

The mummy of the boy king was encased in a nest of three coffins, the inner being most spectacular. It was six feet, one inch long and weighed 2,448 pounds. It was composed of solid gold! As a work of art the coffin is priceless, but the value of the gold making up the third coffin would be, at the current fluctuating price of gold, worth over $10 million!

Although Tutankhamen was a lesser king, the total riches found in his burial chamber are mind-boggling. It is even more difficult then to comprehend the wealth that must have been buried with the great pharaohs such as Ramses II. The treasures that were to accompany him to the next world would probably stagger the imagination.

Scientists recently have been taking a second look at Egyptian tomb robbing to discover what happened to all the treasures buried in the tombs. More gold appears to have been buried with the Egyptian kings than has been accumulated by Western man throughout history. Yet only a minute fraction of the Egyptian treasures has ever been found, since the royal tombs were routinely robbed back in ancient times. Under these circumstances one can't help but wonder how and why each successor to the throne became fabulously rich in a very short time, while his predecessor's tomb was systematically robbed. Was the wealth, as scientists suspect, stealthily passed on?

• •

Population Explosion

If animals overgrazed, overhunted, and exploited sources of food as man does, all living things would have vanished from the earth a long time ago. Worse yet, they could overpopulate, as man does, and accelerate the process of extinction.

Most animals, however, prevent overpopulation by engaging in some type of instinctive behavior. Perhaps the most dedicated practice of birth control is demonstrated by the gannets on the coast of Newfoundland. Here at Cape St. Mary's, several cliffs hang precipitously over the water, but just one is used as a nesting site.

Only the birds that arrive early enough in the spring to win a nesting place on the cliff are allowed to mate, lay eggs, and raise offspring. All others, those that arrived a little late or were displaced by stronger birds, are banished to a neighboring cliff. Here all they do is observe the activities of the many courtships in the nesting sites and wait until the next vacancy or next year.

Since the banished group consists of both males and females, there would seem to be nothing to prevent their building nests and having young. But they absolutely do not. It is as if an invisible borderline surrounded the time-honored mating site on which the gannets have bred for many generations. All birds beyond the accepted breeding grounds appear to be under the spell of an instinctive, very rigid sexual taboo. They follow it implicitly so procreation is suppressed.

The apparent reason for all this lies in the surrounding sea. These waters contain a plentiful supply of fish, the gannet's main food item. The birds seem to know that to avoid exhausting the fish supply by overpopulating they must restrict mating. Only in this manner can they control population growth and avoid probable depletion of the food supply. The gannet thus exerts controls of its numbers before the fish become scarce, thereby assuring adequate sustenance for many future generations.

Scent figures prominently as a birth control device among many animal species. A most impressive example, harsh as it may seem, is that of the meal beetles. These insects inhabit mills and granaries and usually reproduce rapidly. But as soon as their population exceeds two beetles to a gram of flour, the female reacts rapidly by devouring her own eggs the moment they are laid. The trigger for this shocking behavior lies in a chemical substance contained in the beetle's excrement. As the substance increases in concentration, its scent first causes the fertility of the female to diminish and then it prolongs the duration of larval development; the final stage includes the aforementioned egg cannibalism!

In 1859 Thomas Austin turned twenty-four rabbits loose in Australia. Within six years the rabbit population grew to two million, and that was only the beginning of the rabbit explosion. They now overrun Australia, occupying every possible nook. But here's the rub—they no longer "breed like rabbits." Why not? Simply because the fertility of these animals depends on the weather, and it has become a very strong instinctive control.

In times of extreme drought the males resolutely refrain from approaching the female. They seem to be aware that offspring would not survive in such conditions. If the female is already pregnant when the drought sets in, she will miscarry during the hot, dry days that follow.

All is not lost, however, because as soon as the first major rainfall occurs, signaling the end of the drought, the rabbits react almost instantly. What do they do? Naturally they return to their favorite pastime of making babies.

These are just a few examples of the ways animals prevent overpopulation; various species use many other self-regulatory controls. But what happens when irresponsibility is the order of the day?

Such species as lemmings and locusts, and others in which the regulatory system is ineffective, reproduce to a point where other, less palatable systems intervene. In many animals, when population density exceeds a prescribed limit, their social and moral behavior disintegrates. Quarrels break out within groups, offspring are neglected, polygamy and adultery replace monogamous relationships, and the entire society degenerates. With this comes a high mortality rate, and only when the population is stabilized at a reasonable density will the social order return to normal. As man may discover, from overcrowding to total nothingness is really only a step.

Stone Age man of 30,000 years ago lived in widely separated clannish groups, each consisting of about twenty-five people. In all of England there were probably no more than 500 humans. Scientists estimate that in the entire world the population of early man during that period was only around 3.34 million, about the same number of people as live in present-day Chicago!

It took about 30,000 years for the human species to produce a billion people. The same number is now produced in about ten years. At present the human species grows daily by over 217,000 people, about the size of Fresno, California. By the end of a single week the human numbers worldwide will have swelled by the equivalent of the population of New Hampshire and Vermont! But why?

Compared with other species, the human female is rather unproductive—usually fewer than five offspring in a lifetime. Contrast this to the common toad, which produces over 7,000 eggs at a single laying; or the herring, with 50,000 eggs at one time; or the queen termite, which delivers eleven million eggs annually for fifteen years!

One cannot help wondering why the human, with such a low productive capacity, is in great danger of overpopulation. While nature has provided methods of keeping life on earth in an equal balance, her rules don't seem to apply to man.

Some of the restricting conditions on animal population are natural predators, unfavorable environments, the destruction of the natural environment, and restricted opportunities for mating. In contrast the human species suffers no natural predators, controls the environment for its own benefit (usually at the expense of some defenseless animal species), and probably above all, has no physiological restriction in its mating activities.

Overpopulation is a prelude to disaster in all animal societies. However, man is perfecting an old but very effective method of achieving zero population growth. It is a form of mass suicide called war.

There is a tragic ending to this story of population controls, because the self-regulatory method does have its dark side. In many places on our globe animals are exposed defenselessly to the depredations of man. When hunting, farmland expansion, or plain, unadulterated extermination causes a species to fall below a critical limit in population density, disaster falls very suddenly on the remaining stock. The will to live and propagate the species seems to dissipate, and the small number that is left will die out of its own accord. The species will, on a voluntary basis, vanish from the face of the earth—forever.

• •

CHAPTER FOUR

Even Death Does Not Part

The male and female coyote live in an enviably close family relationship. They hunt together, doze on a sunny hillside, and sing duets to the moon. They are most cooperative on the hunt.

Since rabbits can run much faster than coyotes, the latter often hunt in relays. One chases the rabbit toward a specific place, where out pops its partner to chase the rabbit in a new direction. Moving crisscross at a leisurely pace, resting to regain its wind, the first coyote cuts in again. This continues until the rabbit is exhausted and easily caught.

The lifetime commitment in the coyotes' relationship is so compulsive that they may even choose to die together. If one coyote is caught in a trap, its mate will often refuse to abandon it, but will bring it rabbits, squirrels, and other small animals to sustain it. The free mate will even soak in a nearby stream and lie next to the captive so that its trapped mate can chew on the wet fur to relieve its thirst.

Some trappers take ruthless advantage of this loyalty and will leave the trapped coyote alive as bait to attract its distraught mate. Unfortunately this stratagem works all too often.

The suffering of the trapped coyote is difficult to imagine since the traps used are steel-jawed leg-hold devices that crush the animals' bones. In many instances animals caught in such a trap have been known to gnaw off the entrapped limb in order to escape. They frequently die anyway, from either shock or blood loss.

Family loyalty to a trapped animal is hardly restricted to the coyote. One documented case recounts an Alaskan lynx that was caught in such a trap for almost six weeks while members of its family continued to feed it. When the trapper finally showed up to put the trapped animal out of its misery, he was able to include several of the attending relatives in his bounty.

Sperm whales are another group of animals with strong family ties. As parents they are particularly conscientious. A sick or wounded calf may be held above water in the mouth of its parent or may ride on the adult's back until it recovers or dies. As long as a flicker of life remains in the young whale's body, it will not be abandoned.

A harpooned female sperm whale is often supported by the male, who will remain with her even though he himself is in danger of being killed. Early whalers and those of modern times have taken advantage of this family loyalty; it has been a significant factor in reducing the sperm whale population.

What passes through the mind of the person who can fire in spite of the vision of a huge, grieving sea mammal, trying to sustain the life of its mate? Perhaps the gunner asks for mercy as he fires his harpoon gun.

Animal loyalty is also common among some of the feline carnivores; even the solitary leopard at times displays devotion. A hunter in East Africa put out some poisoned bait to exterminate a leopard that had taken to killing cattle. Unfortunately it worked all too well; the next morning he found a female leopard lying dead across the bait. She was not alone, for her mate was beside her, quite alive, with his head laid across her body in a caressing manner. He would let no one near, and nothing the hunter could do would induce the male to leave his dead mate. It almost seemed merciful when he was finally shot.

Not all instances of animal loyalty end in tragedy; occasionally animals are able to rescue their fellow creatures. In a recorded incident, just off the coast of southern Florida, a group of fishermen were being entertained by the frolicking of three large porpoises. The sea mammals seemed to know they were entertaining a human audience. They swam around and around the boat, frequently leaping high into the air as porpoises typically do when following a boat or at play.

A large coral reef grew nearby, a fact that one of the sea creatures seemed to have forgotten. One porpoise, carried away by the increasing intensity of its play, swam headlong, and at great speed, into the solid reef. It must have hit its head violently; it was rendered unconscious and began to sink. Surely the porpoise would have drowned had it not been for the rescue efforts of its two buddies. Apparently they realized their companion was injured and quickly positioned themselves on either side of the unconscious porpoise. They deliberately buoyed it up so that its blowhole was always above the surface of the water, enabling it to breathe.

The fishermen who witnessed the event swore that the porpoises held their unconscious companion above water for at least fifteen minutes. They all began to cheer when it was evident the injured porpoise was regaining consciousness. Some insisted that, when it began to move, the rescuing sea mammals swished it through the water as if to hurry its return to the living.

In a short period of time the three porpoises were again frolicking and leaping from the water, but only for a few minutes before swimming away. Very likely this after-incident play was cut short because one of the three had a giant-sized porpoise-type headache.

• •

An Arctic Dinosaur

On August 3, 1960, a party of geologists exploring a sandstone cliff made a most astonishing discovery. Exposed to view was a series of thirteen footprints of a large dinosaur later identified as a species of *Iguanodon*. The prints were thirty inches long, unusually large compared to other *Iguanodon* tracks found in Europe. They were made during the last period of the Age of Reptiles, the Cretaceous, about one hundred million years ago.

The most incredible feature of this discovery was that it took place in Spitsbergen, well within the Arctic Circle. Enough is known about dinosaur environments to conclude that, with few exceptions, all species lived in tropical or subtropical regions. A dinosaur of this size would certainly require much vegetation to sustain itself. An Arctic home, such as Spitsbergen, could not possibly produce enough plant food to support herds of such large animals. Yet here, within the Arctic wasteland, were dinosaur footprints! The obvious conclusion is that Spitsbergen could not have been in a frigid climatic region when *Iguanodon* walked across its surface.

There is much geological evidence to indicate that during the days of the dinosaur most of the present continents were united into a supercontinent now referred to as Pangaea. The breakup in the land masses we know today took place during the final millennia of the dinosaur era. In time the various lands "drifted" to their present locations, a process referred to as *continental drift* or *plate tectonics*.

The discovery of *Iguanodon* tracks has added indisputable evidence to the theory of continental drift. The presence of such a dinosaurian species clearly indicates that one hundred million years ago Spitsbergen was tropical. Therefore, it could not possibly have been in its present geographic location at that time.

Spitsbergen must have been part of Europe during the time of the dinosaurs. Geologists know that much of Eurasia at that time was tropical. During the breakup of the continental masses Spitsbergen split off from Europe and, in the course of millions of years, drifted to its present position.

• •

The Saga of Hubert the Hippo

The hippopotamus can be a dangerous and unpredictable animal, especially when irritated. However, most Africans regard this animal with no more alarm than would a Western motorist braking for livestock. Many Africans look on the hippo with fondness, and some have even become folk heroes. The most celebrated was a large hippo affectionately known as Hubert.

Most hippos rarely wander more than a half day's walk from their home waters, but not Hubert. Early in the 1940s he set out on a 1,000-mile trek that took him through drought-stricken South Africa.

At the very outset of his long walk it was quickly observed by backcountry villagers that rain often followed his arrival in the area. He was soon warmly welcomed and even worshiped as a special spirit sent to relieve the drought. When he entered more settled regions, local radio stations and newspapers began to report regularly on his progress. His arrival in a city usually took on the aspect of a parade. Children ran alongside the huge wanderer, while adults held out food and cheered him. Most astonishingly, Hubert seemed to be aware of what was going on.

He once tried to enter a theater where a Judy Garland movie was playing, but was ushered out by the manager. Doubtless he was convinced that none of the seats would support the visiting hippo.

By now Hubert was a true vagabond, wandering casually along highways or city streets and grazing calmly in parks. Newspapers and radios continued to report daily on the whereabouts and doings of this strange nomad.

Hubert's journey was not to continue indefinitely since all things, good or bad, must come to an end, however ignobly. All of South Africa was saddened by his death, with one possible exception. Hubert had stopped to have lunch in a large vegetable garden. The irate landowner did not approve and promptly shot him. Thus ended the saga and wanderings of Hubert the Hippo.

• •

The Colossus of Rhodes

Over 2,000 years before the Statue of Liberty was built in New York Harbor, the inspiration for the modern monument had been created and destroyed. The original model, one of the Seven Wonders of the Ancient World, was the colossus of Rhodes.

The island of Rhodes boasted many lofty images of the gods, but by far the greatest was the colossus dedicated to the sun god Helios. It was a hollow bronze statue that stood well over one hundred feet tall. According to legend, ships sailed between the legs of the gigantic statue as it stood astride the harbor.

Prior to the building of the statue, Rhodes had been besieged by Demetrius, a Macedonian king. Although the assault was devastating, he failed to capture the city and finally abandoned the siege. Demetrius left behind huge conglomerations of equipment that, by modern standards, would be worth millions of dollars. It was the sale of this booty that financed the building of the colossus.

Twelve years were spent in completing the statue (292-280 B.C.), and for the next fifty years the colossus served as a beacon for all incoming ships. The gigantic figure must have seemed eternal to the citizens of Rhodes. But the forces that stir the earth had different plans; about fifty years after it was completed, a major earthquake shook the area with such violence that the colossus fell.

So tremendous was the impact that the giant bronze statue shattered as it struck, and there the fragments lay, almost completely undisturbed, for nearly 1,000 years. Even in its dismantled state people continued to marvel at the enormous wreck, which served as an ancient tourist attraction.

Three centuries after it fell, Pliny, the Roman historian, wrote about the colossus, ". . . But even as it lies, it incites our wonder and admiration. Few men can clasp the thumb in their arms. . . . Where the limbs are broken asunder, vast caverns are seen yawning in the interior."

The fragments of the statue were finally sold to the Saracens in the year A.D. 653. According to reports, 900 camel loads of bronze were carried away. Ironically, the materials were put to the same use the Macedonians had employed them for 1,000 years earlier—making implements of war!

• •

The Lake of Gold

The American West abounds in legends of buried treasure and fabulous lost mines. Most are nothing more than tall tales based on overactive imaginations, but a few of the legends have some basis in fact. . . . This is one of them.

In the year 1849 Richard Stoddard and a companion became lost while prospecting somewhere in the vast Sierra Nevada in California. By chance they stumbled onto a small lake and, as they stooped to drink, suddenly forgot their thirst. The shoreline of the lake was literally covered with nuggets of gold gleaming in the sunlight. Taking many samples, the two prospectors tried to make it back to civilization. Their hardships were many, and only Stoddard survived, his partner having been killed by Indians along the way.

It was a human wreck who stumbled into Sacramento many weeks later. There, during the next few months, Stoddard was gradually nursed back to health. He still had on his person a number of nuggets, but sick of mind and spirit, he had no memory of how they had come into his possession. Eventually, as his health returned, so did his memory—more or less.

His tale of a golden lake spread quickly—almost frantically— for he had a fistful of nuggets to verify its existence. Stoddard had no difficulty assembling a group of miners who were willing to finance the expedition and return into the mountains with him. Each was to have an equal share.

Weeks later, as they wandered deep into the great Sierra, it became quite evident to the miners that Stoddard was not sure where he was. Day followed day, and frustration followed frustration. Finally, angered by his futile attempts to relocate the treasure lake, the miners gave him just twenty-four hours for one last chance at success—or he would be hanged!

That night, under the cover of darkness, Stoddard slipped away and was never seen again. The area, now known as Last Chance Valley, is indicated on local maps. It serves as a guide for those who would seek the lake whose shores are strewn with gold gleaming in the midday sun!

• •

A Sense of Timing

Some animals seem to possess a remarkable sense of timing; foremost among them are the bees. In 1921, the entomologist Karl von Frisch performed an experiment to determine whether bees can be attracted to the same spot at a precise time each day. At a designated time and place he laid out foods that would attract bees. The bees visiting this outdoor cafeteria were caught and marked with tiny spots of red paint. The conditioning of the bees to respond to the stimulus continued for some time, during which they acquired the habit of arriving regularly for the free meals.

When a timekeeper was stationed to record the bees' arrival, he found the majority arrived at almost precisely the time the food was regularly set out. When no food was set out, the bees arrived at their usual mealtime anyway. When the food was withheld for a week, the bees continued to return promptly at the time they had learned to associate with the food. After that their numbers began to dwindle; a few continued to arrive on schedule for several weeks.

An example of even more precise animal timing is illustrated by the Australian fruit bat. These animals congregate in vast numbers among densely foliaged trees. Promptly at six o'clock every evening the bats stream out to wherever their feeding grounds are located. Whether it's daylight or after dark, summer or winter, they always appear at the same exact hour. Australians claim that the punctuality of this daily maneuver is so precise that they can set their clocks by the time of the bats' evening flight.

• •

The Taming of Man's Best Friend

The dog was probably the first animal species to be domesticated by man. Its wolf ancestors hunted throughout the prehistoric forests, enemy to all they encountered. But somewhere, about 50,000 years ago, they discovered man. And so began the taming of man's best friend.

Man's earliest association with the carnivorous canine breed, which would in time become his best friend, was very likely not a happy one. Our brutish ancestors of about 250,000 years ago placed food and self-preservation at the top of their survival list. Stealthily the cave dweller must have followed a bitch, her pendulous teats identifying her as a nursing mother, to her lair. Here he split her skull with a stone ax and applied the same treatment to her whelps. These were delicacies that he brought back to his lair to feed the hungry family awaiting his return. It is not difficult to envision the same savage Stone Age man being stalked and pulled down by the slavering jaws of a hunting pack of primeval dogs.

It seems unlikely that humans of the Old Stone Age, prior to 50,000 years ago, had any pets. The doglike wolves they knew were at first discouraged from approaching the encampments.

At the dawn of mankind, no social etiquette restrained the genus *Homo*, and early man doubtless lived in an aura of animal filth. In those days cleanliness was certainly not one of the basic human needs. Therefore the outside of his cave, and probably the inside to a somewhat lesser extent, was a receptacle for all manner of human refuse and garbage. Sooner or later a wolf-dog, disabled or too old to hunt, must have been stimulated by the scent of rotting food and come to man's cave, furtively carrying away a bone the humans had tossed aside. Doubtless other members of its pack followed, partaking of the accessible sustenance left outside the caves. None were welcome, but the bolder wolves continued to lurk near the outskirts, foraging for scraps of meat or discarded bones. As time passed, they were gradually tolerated and allowed to serve as scavengers until eventually they became part of the settlement. The wolves in return took up the role of watchdogs, sounding a bark at the approach of any strange beast.

There doubtlessly came a time when a hunter, following a bitch to her lair, would save a pup from the litter and bring it back to his cave for his own offspring to play with. And so man and dog began to dwell together for the first time under the same overhang.

The men soon discovered the value of the dog, with its agility and keen scent, in the hunt. A pack of tame dogs could chase a deer into exhaustion so the hunters could close in. Their reward: good pickings after the kill.

As scavengers, watchdogs, and hunting companions, dogs gradually acquired a permanent role in human society. The remains of every human encampment since 10,000 B.C. show the presence of the domestic dog. And to many children of those days long ago, the dog was a playmate and an inseparable companion. A recent excavation of a Paleolithic site in Israel yielded the skeleton of a boy flexed on his right side. The boy's left hand rested gently on the remains of a three- to five-month-old puppy.

Such burials are not unusual for ancient man. Many scientists believe that by the Late Stone Age dogs may have already been selectively bred by the early hunter-gatherers. Watchdogs, hunters, and pets were then joined by pack and herding types.

The changes in the now domesticated dog had only begun. As dogs began to accompany migrating people to other areas of the world, they would be preselected or bred to fit the new environment, and their work was cut out for them. To guard the palace or temple, guide the infirm, tend flocks, raise an alarm, fight in the arena, attack the enemy, carry messages and defend the camp in warfare, collect fleas from milady, warm worshipers' feet—those are but a few of the tasks that dogs through the ages have learned to handle capably.

It is no wonder that during the last 10,000 years dogs have become increasingly diverse. From the toy poodle that fits into a teacup to the Irish wolfhound and from the rare and valuable Chinese Shar Pei to the "mutt," dogs' most important role is to be good company. Only a small number of these domestic animals even resemble that common ancestor of all dogs—the wolf.

Shades of the Ancient Mariner

For many centuries sailors have believed the albatross to be a reincarnated mariner following ships to warn of danger and storms. To kill one would bring eternal bad luck. The legend was detailed graphically in Samuel Taylor Coleridge's famous epic poem, "Rime of the Ancient Mariner," written in 1798.

As recently as 1959, when a ship developed chronic engine trouble during a voyage, belief in this legend resurfaced. When the vessel limped into port for repairs, the crew deserted the ship en masse. The reason? The ship was carrying a caged albatross, and the men believed this abuse of the

bird was causing the bad luck the ship was experiencing. They anticipated a major disaster if they were to put out to sea again.

The superstition surrounding the albatross also played a role in an incident that occurred during World War II. When a scout bomber plane was downed in the Pacific in 1942, the crew survived and drifted in the open sea on a large raft. On the edge of starvation, they managed to shoot an albatross, which they hauled aboard. They skinned it and ate the internal organs, setting the remainder aside.

During the night one of the men awoke and became possessed by supernatural terror. The remains of the bird were glowing so brightly that the raft and the sea around it were brightly illuminated. The fright of the man caused him to suddenly remember the saga of the ancient mariner His screams quickly awakened his companions. Although they were starving and had no prospects of finding any additional food, they immediately threw the remains of the bird overboard. Later, when they were rescued, they learned the skin of the albatross is often highly phosphorescent because the species preys primarily on luminescent fish. The phosphorus is absorbed by the skin of the bird, so when its feathers are removed, it glows in the dark.

• •

"Come into My Parlor"

The manner in which predators bring down their victims often seems cruel and inhumane. It must be remembered, however, that the intent of the predator is always to subdue its prey as quickly as possible. It matters little whether or not the carnivore's purposes are altruistic, because the speedier and more efficient the kill, the more compassionate it appears.

The process through which prey animals become food for carnivores is extremely important in maintaining a balanced ecosystem. In almost every habitat where the predator has been depleted, usually by man, the natural prey multiplies beyond control, and many die of starvation because there is not enough food to go around. On the other hand, if predators killed unnecessarily in the wild, they would soon run out of food and precipitate their own destruction.

The lion, the tiger, and the leopard are prize examples of talent and skill in hunting, but the most accomplished hunter of them all is the lowly spider. The great cats, which fail to capture their prey many more times than they succeed, could indeed learn a few tricks from the modest-sized arachnid.

Much of the spider's efficiency in hunting is the result of advance planning. Basic to the spider's predatory proclivity is its instinctive, quick, and accurate skill in web making. It may be an untidy sheet of web, a silken tube, several sticky strands, or an intricate geometric orb, sometimes six-and-a-half feet in diameter, containing up to 1,000 feet of silk. The spider will produce whatever type of web serves it best as home, storage area, camouflage or decoy, and, of course, trap.

Beyond its skill of building a web in which to lie and wait for victims, many spiders exhibit special techniques that warrant the admiration of every creature except their unfortunate prey. There is the spitting spider, which fires a sticky thread from its jaws to pin down its prey. The bola, or angling spider, produces a weighted thread, or fishing line, laced with blobs of sticky gum. When it feels the vibration from a potential captive, it whirls the line and reels in whatever luckless prey has made contact with a globule of gum.

The water spider, finding its prey in lakes, ponds, and ditches, fashions a diving bell out of tightly woven silk, into which it carries a supply of air bubbles. The Mediterranean orb spider has perfected a defensive tactic by which it arranges the carcasses of insects in two piles that match the shape and color of its own body. The decoys are placed strategically on the web so that an enemy has only one chance in three of catching the real thing. While the predator is concentrating on the decoy, the spider works out a strategic retreat. Each spider has refined its skills so expertly that by the time the spider needs a meal all that remains to be done is to start eating.

Over 40,000 known species of spiders are living today. The spider is largely responsible for controlling the insect population beyond the scope of spray, swatter, and exterminator. So well adapted is this predator to killing and devouring its prey that, in England and Wales alone, the number of insects consumed by spiders in a single year is greater than that of the human inhabitants of these countries combined.

There is no doubt that the spider is indeed the greatest of hunters and one on whom we have grown to depend. The arachnid is often unaware of this, as an incident in Honduras might illustrate. The workers on the banana plantations complained of an alarming overabundance of spiders, and this was undermining their efficiency, to say nothing of their job contentment. The owners of the plantations, yielding to the workers' complaints, finally had the area sprayed extensively, and the spiders were exterminated.

The natural balance between predator and prey was definitely disturbed, because the next crop was consumed almost entirely by insects that were resistant to spray chemicals! The banana companies quickly imported a supply of spiders to get them out of the ecological mess they had created. No further complaints from employees were recorded.

• •

Souvenir of Civilization

A noted exploration geologist recently remarked that on many occasions he had been in such remote desert country that he was convinced that nobody had ever been there before. His conclusion was almost always belied by his stumbling on such relics of modern life as discarded garbage.

The discovery of such human artifacts on all corners of the planet has indeed reached ridiculous proportions. When Admiral R. J. Galanson, chief of the United States Naval Materials, peered though the porthole of a navy deep-submersible craft, he was almost a half mile below the surface of the Pacific Ocean.

With his adrenaline working overtime, the admiral truly expected to see sights no human being had ever seen before. He was quickly made aware of his error as he surveyed the ocean floor—the very first thing to meet his gaze was an empty beer can!

• •

The Crab Nebula

Supernovas are cataclysmic star explosions during which a star may increase in brightness by a factor of a million or more. The star literally blows itself apart; more than half of its mass is lost to space, and the star retains very little of its former nature. Supernovas are extremely rare. Astronomers estimate that in any given galaxy there are only three occurrences in 1,000 years.

Within just a few days after a supernova occurs, the exploding star brightens by many thousands or millions of times. At its brightest a supernova can rival the combined light output of all the stars in a galaxy such as the Milky Way. At least half the star's original mass is thrown off at speeds of up to 6,000 miles per second. The core of the destroyed star, without the internal energy source of its nuclear fires to sustain it, collapses to form a tiny compressed star, smaller and denser than a white dwarf. The strong inward pull due to the tremendous gravity of the star's core, aided by the tremendous pressures of the supernova explosion in the layers above it, crushes together the electrons and protons of the core's atoms, forming electrically neutral particles called *neutrons*. Scientists refer to such a celestial object as a *neutron star*.

A neutron star could contain as much matter as two of our suns compressed into a single sphere no more than twelve miles across. The density of such an object is so great that a thimbleful of material from a neutron star would weigh at least 1,000 million tons.

The presence of neutron stars was theorized as long ago as 1939, but

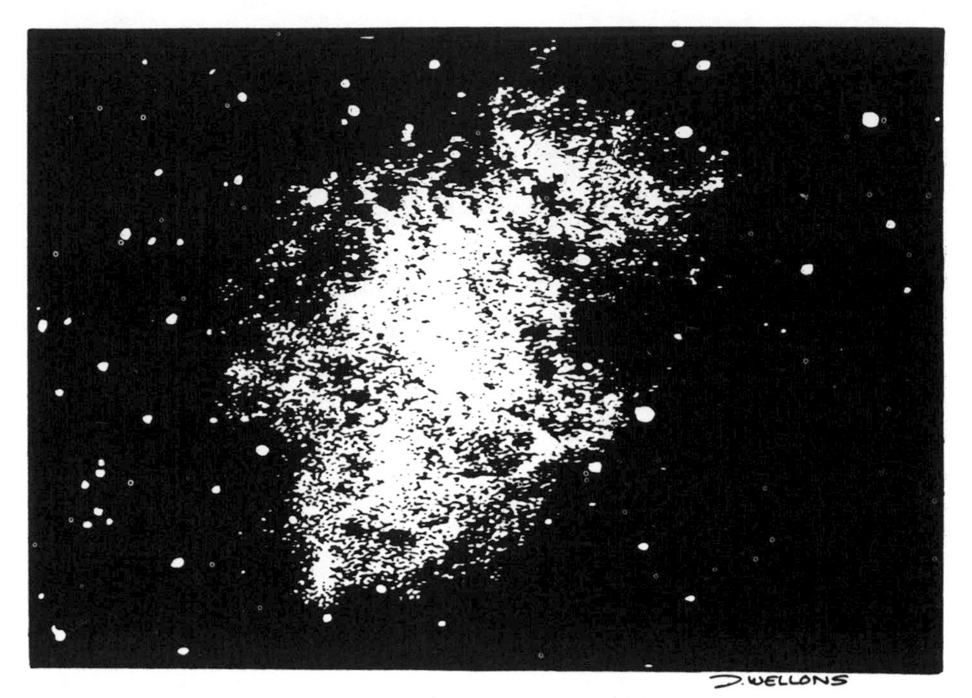

there seemed little chance of detecting them. In August 1967 astronomers at Cambridge University made recordings of regularly beating radio waves coming from a star. It soon became clear that it was a tiny star that rotated rapidly, sending out shafts of energy each time it turned. Such stars are now referred to as *pulsars*.

Approximately 150 pulsars are now known to astronomers. The periods of their pulses range from thirty per second to one every four seconds. The fastest-flashing pulsar lies in the heart of the Crab Nebula, the collapsed core of a star that exploded in 1054. Its presence in that nebula underscores the theory that pulsars and neutron stars are in reality the same thing. Scientists generally agree that pulsars are small, rapidly spinning neutron stars that radiate energy in pulses of electromagnetic radiation.

In the astronomically fateful year of 1054 Chinese astronomers recorded the explosion of a star, a supernova, in the constellation of Taurus. Since it appeared suddenly out of nowhere, they referred to it as "the guest star." The guest star shone as brightly as the planet Venus and was easily seen in the daylight for the next twenty-three days. It remained visible after dark for over 650 days before finally fading from view.

The star certainly must have been visible in Europe and the Middle East as well, but no account of it exists. While these advanced civilizations inexplicably ignored one of the more spectacular phenomena of the universe—the creation of the Crab Nebula—prehistoric Native Americans did not.

Scientists believe that two prehistoric Indian rock drawings found in the early 1950s in northern Arizona depict the event. The star appears near a crescent moon, the position in which it would have been seen in 1054. Since then, nine more rock drawings have been found, all showing the same event.

The remains of the supernova, now known as the Crab Nebula, consist of dust and gas that expand, due to the momentum of the original explosion, at the rate of 800 miles per second. After nine centuries of travel the Crab Nebula is now more than forty-two light-years across!

At present the nebula is a source of strong radio waves. Two faint stars are at the center, one of which, the neutron star, is the collapsed core of the star that exploded. As small as it is, this center star is bright enough to show up on photographic plates, and scientists have long suspected that it was the remnant of a supernova explosion. In 1968, once astronomers knew what to look for, they found that this star was flashing optically thirty times per second—the same rate as that of the radio waves they had been detecting for some time.

Only one other pulsar in the southern constellation has been detected optically. It is the star Vela, which is the faintest star ever detected. Vela is ten million times too dim to be seen with the naked eye.

The final outcome of the supernova that created the Crab Nebula can be witnessed in the remnant of another exploded star, the Veil Nebula in Cygnus. The tattered remains of this star, which exploded some 20,000 years ago, are wisps of gas dispersing in space. In about 100,000 years they will disappear from the sight of earthbound telescopes.

• •

A Diet of Dinosaurs

Stories of human encounters with enormous snakes have been passed down through the ages, and like most oft-told tales, they have been subject to some exaggeration. Typically, the reptiles are described as being thirty to forty feet long, and although pythons or anacondas of over twenty feet are not uncommon, a thirty-foot snake is extremely rare. When a snake of such length is reported, evidence rarely exists to substantiate the description.

To date the longest snake on record is a reticulated python with a verified length of thirty-three feet. No known anaconda has reached that length, but stretched skins often accompany inflated stories. A standing reward of $5,000 has been posted for anyone who can prove the existence of the often reported forty-five-foot anacondas. To date nobody has collected the reward, and probably no one ever will.

The only foolproof way to find the ultimate big snake would be to go back in time sixty to one hundred million years. Scientists recently uncov-

ered fossil remains of a sixty-million-year-old python, *Gigantophis,* in El Faiyûm, Egypt. Enough of the snake was recovered to enable the scientists to estimate its length rather accurately. All agreed that in life it must have been at least sixty-five feet long.

And as enormous as this may seem, fossil remains of an even larger species of python are known to have coexisted with dinosaurs millions of years before the time of *Gigantophis*. Such stupendous size should, of course, be expected of serpents of one hundred million years ago—after all, they dined on dinosaurs.

Fish Stories

When people refer to small fish they usually identify them as minnows, as though this were generic for any common small freshwater fish. But in the western United States, particularly in central California, dwells a fish called the Colorado squawfish, which frequently weighs as much as one hundred pounds and grows to nearly six feet in length. It may be disillusioning to fishermen, but the squawfish is a minnow. Not surprisingly, it is the largest minnow in the Western Hemisphere!

No doubt many humans have great affection for their pets and would go to any length to keep them healthy and happy. But even the care and pampering of a pet can be overdone. A recorded case in Newfoundland tells of a man whose unreasonable fondness for his pet fish cost him his wife. The woman applied for a divorce on the grounds that her husband insisted on keeping a large catfish in the only bathtub in the house! The divorce was granted.

A husband bent on going fishing, over the protestations of his nagging wife, is a classical theme depicted in television situation comedies. She is the harping nuisance who behaves unreasonably, gets in the way, and sabotages the entire fishing expedition. It may be funny, but it is far from reality, for there is no reason why the wife could not be equally skilled at fishing, as demonstrated by a most unusual fish story.

Recently, while a couple was angling in a North Carolina lake, the woman managed to snag her hook. Her noble husband scrambled into the water to free her line and found it held fast to an automobile tire. He threw the tire up onto the bank to remove the hook, and there ended the day's fishing venture—out of the old tire came nine pan-ready catfish!

A fisherman recently managed to hook a fair-sized bass in a small lake in Illinois. There was nothing particularly unusual about this feat except that the fish was wearing spectacles! Obviously someone must have dropped the glasses overboard, and they subsequently were caught in the fish's gills, This caused them to hang partially over its eyes. On the other hand, perhaps the fish was simply trying to get a better look at what was on the other end of the line.

• •

Ancient Guide Birds

Long before the European discovery of the New World, seafaring men navigating uncharted waters used birds to determine in which direction land lay. To ensure that they got this information they used only birds that could not swim. When released, the birds would return to the ship if they were unable to find land; if they did not return, the sailors knew that land was not far off and they would then sail in the direction the birds had flown.

Many scientists believe it was in this manner that the ancient Polynesians were able to settle the numerous tiny islands in the vast Pacific—centuries before Europeans even dared to venture into the open sea. There are many legends of such events. The early Hindu legends frequently tell about seafaring merchants who carried land-dwelling birds with them to serve as their guides to the nearest shore. And there was Noah, whose dove, on her third trip from the Ark, "returned not again unto him anymore."

The Vikings, who were adventurous seamen, frequently used birds to guide them to land. This may even have been the means by which Leif Eriksson was led to the great land he called Vinland, centuries before Columbus discovered the New World.

• •

The Great Ice Age

Glaciers have moved across the face of the earth many times in the geologic past, always sculpting the land and leaving behind numerous lakes and swamps. The oldest known glacial epoch occurred nearly two billion years ago. In southern Canada, extending about 1,000 miles from east to west, is a series of deposits of glacial origin. This earliest Canadian ice sheet would have been at least 1,000 miles long, 1,000 miles wide, and doubtless thousands of feet thick. New evidence indicates that the ice flowed into the northern United States, at least into the area now known as Michigan.

Glacial deposits found in South Africa, central India, and western Australia are thought to be of an age equivalent to those in Canada. If so, the Ice Age of two billion years ago must have been quite extensive and probably lasted for many millions of years. Since then the earth has undergone many periods of refrigeration.

The cause of these periodic ice ages is a deep enigma of earth history. Scientists have advanced many theories ranging from changing ocean currents to sunspot cycles. No single theory is sound, and doubtless many factors are involved. One fact that seems certain, however, is that the earth is still in a glacial age.

Within the last billion years or so the earth has experienced at least six major phases of refrigeration that apparently occurred at intervals of about 150 million years. Each may have lasted as long as fifty million years.

The term *ice age* can be confusing, since it generally refers to a period when parts of the earth are undergoing glaciation for a period lasting one or two million years. These epochs are usually marked by a series of glacial advances and retreats (interglacials). But when geologists refer to a period

of global cooling that can last for many millions of years, they are talking about an *ice era*. The fifty-million-year epochs mentioned above are classified as ice eras. Understandably then, several ice ages can and do occur within an era.

The most recent ice era began about sixty-five million years ago. It was slow in starting, and its effects were mild at first. But approximately fifty-five million years ago glaciers began to form in Antarctica. The ice grew, shrank, grew again, and then gradually expanded until it coalesced into the dome-shaped ice sheet that, by twenty million years ago, covered the entire continent, as it still does. It wasn't until about twelve million years ago that glaciers began to spread in the mountains of Alaska. Greenland's glaciers are relatively new within this ice era, because this land was not covered by ice until three million years ago.

Now the stage is set for the great Ice Age, as many refer to it, a time that geologists call the Pleistocene. Nearly two million years ago a series of ice advances began, at times covering over one-fourth of the earth's land surface with great sheets of ice thousands of feet thick. During this last epoch of refrigeration the ice advanced and retreated by melting at least four times. Evidence appears to indicate that each succeeding glacial advance was more severe than the previous one. The most severe began about 50,000 years ago and ended about 10,000 years ago.

It is interesting to note that, during the times of glacial retreat, the interglacial, worldwide climates became on the average much warmer than they are at present! Usually the interglacial phase lasted many thousands of years.

Ever since the climax of the last advance, the ice has been in a stage of retreating, and world climates, although fluctuating, are slowly warming. Scientists consider the earth still to be in a glacial stage because one-tenth of the globe's surface is still covered by glacial ice.

Greenland and Antarctica are capped by five million cubic miles of ice, and valley glaciers are common in the mountains of the world. However, well-kept records clearly show that the last hundred years have seen a marked worldwide retreat of ice. Swiss resorts built during the early 1900s to offer scenic views of glaciers now have no ice in sight! If this glacial retreat continues, and all the ice melts, sea level would rise 200 to 300 feet, flooding many of the world's major cities. New York and Boston would then be visited only by scuba divers.

Perhaps the retreat is only temporary, and thousands of years in the future the earth will cool and undergo another period of refrigeration. Huge mountains of ice will then re-form and advance on the land, engulfing whatever civilization stands in their path. Or perhaps the Ice Age is really ending; only time and submerged coastal cities will tell.

• •

Pearls—Gems of the Ages

Pearls have been revered and treasured from the very earliest of times. In fact they were considered the most valuable of gems until late in the nineteenth century, when diamonds supplanted them. It is not difficult to understand the attraction pearls held for ancient peoples. Their translucent beauty embodied such virtues as purity and chastity. Religious symbolism was also eventually added to the folklore of pearls.

Pearls were even used for medicinal purposes. Concoctions of ground-up pearls were drunk greedily by Mogul emperors in the belief that their virility would improve. Effective or not, they probably swore by this preparation. There were other therapeutic uses, all just as bizarre and irrational. Charles VI of France regularly drank concoctions of ground-up pearls in a vain attempt to restore his sanity. Perhaps his mind was too unbalanced to notice the stomachaches that followed his drink.

When Rome was in its glory, no self-respecting Roman woman was without her pearls, for sumptuous pearl jewelry was an ingredient in the Romans' addiction to extravagant luxury. The women not only wore pearls during the day but also routinely adorned themselves with pearls before going to sleep, to ensure that their dreams would be filled with lustrous gems.

The esteem for pearls among the Romans was so extreme that they may have influenced history. It has been reported that Julius Caesar undertook the invasion of the British Isles in 55 B.C. because of the rumors of fine and plentiful freshwater pearls in Scotland. Appreciation for the gems among noblemen rivaled that of the women. Caligula decorated his slippers with

pearls and even draped a pearl necklace around the neck of his beloved horse Incitatus. Nero's scepter was heavily laden with pearls.

Considering how difficult it must have been for ancient humans to obtain pearls, their reverence for this beautiful gemstone is quite understandable. The oldest surviving pearl necklace dates back to about 350 B.C. It was unearthed at Susa in southwestern Iran, the site of a Persian king's winter palace.

Precious gems often yield strange stories of lost mines and buried treasure. Pearls are no exception. Recently an archaeologist, exploring the ruins of a 400-year-old Spanish village, noticed a piece of dried cloth protruding from the ground. His curiosity aroused, he investigated. His excavation brought to light a fabulous cache of over 3,000 precious pearls, all of which had been buried only a few inches under the surface. The piece of cloth sticking out of the ground, which had served as a waybill to this treasure, was one of the few bits of wrapping material that had not disintegrated. Despite this exposure and the 400 years they had spent in the ground, the pearls had lost none of their luster!

The estimated value of the treasure was nearly $500,000. However, the pearls remain shrouded in mystery: who put them there, where did they originate, and why were they abandoned?

• •

"And the Rockets' Red Glare . . ."

Americans unknowingly owe an enduring debt of gratitude to an Englishman for our national anthem. The rockets that so inspired Francis Scott Key as he watched the bombardment of Fort McHenry were the invention of Sir William Congreve, who was, at the time, the royal fire master to the king. During the War of 1812, Francis Scott Key, an American lawyer, was negotiating with the British authorities for the release of certain incarcerated friends. The release was agreed on, but on the night of September 13, 1814, Key was detained aboard an English ship where he couldn't help witnessing the bombardment of Fort McHenry.

The rockets Key observed bursting on the fort were the first ever seen in America. They were made of narrow wooden tubes filled with gunpowder and tipped with iron warheads. The rockets were guided by simple polelike rudders and launched from rows of tilted frames in a series of giant assaults. These early rockets had a range of about two miles and were designed to explode on impact, throwing out a deadly shower of shrapnel from which there was little defense.

As the rockets streaked across the sky with tails hissing and blazing, they must have been a terrifying sight, especially since the assault was so unexpected. However, these missiles did not win the war for England. Instead Congreve's rockets are well remembered today only because their brilliant "red glare" gave America, the enemy, "The Star-Spangled Banner"!

• •

"A Tot of Pot"

Following World War II scuba diving was used in the investigation of submarine archaeology. By use of this form of exploration many ancient wrecks, mostly cargo ships, have been found. It was nevertheless of considerable surprise to one group of divers that their discovery, under only eight feet of water, was a Carthaginian warship. The ship, which had been rammed and sunk in shallow waters off Sicily, was a casualty from the First Punic War, 264-241 B.C. The ramming impact was so powerful that the ship's stern had been driven into the hard, sandy bottom near the shore. Since the water was shallow, the prow of the ship must have protruded above the surface after sinking. A protective covering of sand kept the wooden portions of the ship from decaying for over twenty-two centuries.

The discovery of a warship was itself unusual because the deck would have been kept clear for action and there would be no solid cargo to show evidence of wreckage after the hull was flattened and buried. Ballast stones were found, it appears that most of the crewmen seem to have escaped by weight in the absence of cargo.

The wreck, identified as "The Punic Ship" by the discovery team, revealed much information about the ships themselves and about what life was like for the men on board. Some forty feet of keel, portions of both port and starboard, and the sternpost were discovered, from which the shape and size of the original ship could be calculated. It could carry sixty-eight oarsmen (two per oar) and any number of fighting men.

Fragments of human bones and those of a small dog were found in the bottom of the hull. Ballast stones had tumbled on them, preventing the remains from being dispersed. Because no remnants of battle equipment were found, it appears that most of the crew men seem to have escaped by scrambling ashore with their weapons.

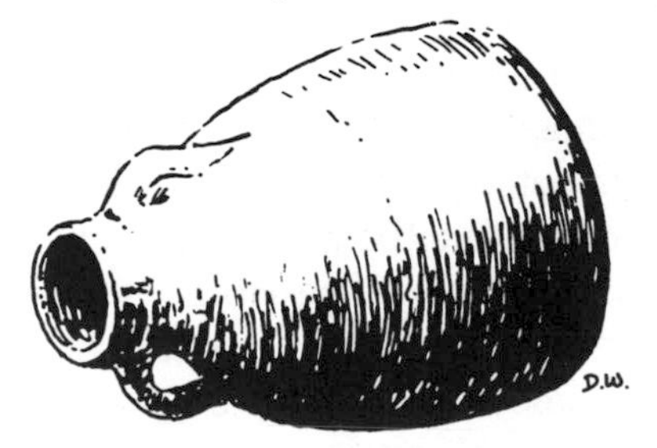

The store of food and crockery for the crew remained with the wreck until its discovery. The crocks were small, suitable for individual quick snacks. Also included among the artifacts were some amphorae, containers that usually carried wine or water. The remnants of food seem to indicate that the soldiers grappled for their rations like fighting carnivores. Butchers had trimmed meat from bones of a variety of animals, and it was apparently served to the soldiers either raw or, at the most, slightly singed.

Surprisingly, the most abundant find was a number of baskets of yellow grass. There was so much of this plant material that a bagful was easily obtained, more than enough for laboratory analysis. The results confirmed that all of the material was *Cannabis sativa*, known today as marijuana and referred to more popularly as "grass," "pot," "weed," "Mary Jane," and a host of other terms.

What can one conclude from abundant supplies of marijuana aboard a fighting ship? Perhaps the soldiers were encouraged to use the marijuana to become euphoric, to reach what is today known as a "high" in order to intensify their courage and fearlessness. The soldiers may have chewed on the stems and infused the weed into a tea. It is well known that in the past the British navy commonly issued a "tot of rum" to its seamen. With the same purpose in mind, the Carthaginian navy must have dispensed a "tot of pot" to its soldiers as they prepared for battle. Small wonder they lost!

• •

Seagoing Snakes

When the subject of snakes is raised, one immediately thinks of deserts, swamps, and jungles. Snakes do occur in these places, but the locale of their greatest abundance is the sea.

There are approximately fifty species of sea snakes, all members of the family Hydrophidae, that roam the waters of the Indo-Pacific region. Like so many groups of animals, there is always one type that is the wanderer. The habitat of one species, *Aelamis platurus*, extends into the western coastal waters of Central America and Mexico in considerable numbers. Its full range is from eastern Africa to western America, making it the most widely distributed reptile in the world. A few specimens have even been observed in Posieta Bay, Siberia.

Sea snakes do not seem able to survive in waters colder than 68 degrees Fahrenheit, so water temperatures form an invisible barrier, as effective as any material wall, that prevent them from entering the Atlantic Ocean around the Cape of Good Hope or Cape Horn.

North and South America were two separate continents prior to three million years ago. It was about then that the isthmus of Panama rose from

the sea, joining the two continents and forming an effective land barrier that would separate the two great oceans. Scientists believe that before that time the sea snakes had not yet inhabited the eastern edge of the Pacific. By the time their migrations finally took them there, the isthmus of Panama was already dry land and served as a barricade. Were it not for that wide stretch of land, the sea snake would certainly inhabit the Atlantic today.

One might wonder why the snakes do not cross the isthmus through the Panama Canal. For them this would be impossible because freshwater lakes in the canal bar the passage of such marine organisms from one ocean to another. Sea snakes cannot enter fresh water because they would quickly lose critical amounts of body salts and die. Normally, salt from marine waters replaces the salt that is continuously eliminated by a special gland located in the sea snake's lower jaw. In this manner the intake and elimination of salt in the snake's body are in balance.

Sea snakes are very highly specialized marine reptiles. They must come to the surface to breathe but can dive to considerable depths, remaining underwater for up to eight hours without breathing, an absolute record among air-breathing vertebrates. Their bodies are well adapted for such a feat because their right lung alone takes up most of the internal body space and extends even into the tail region. Their windpipe also has been modified into an auxiliary lung that, like a true lung, takes oxygen into the blood. Thus the snake uses its intake of air most efficiently.

These seagoing reptiles are generally coastal animals with a distinct preference for the mouths of rivers, harbors, beach areas, and coral reefs. The many species share one common characteristic—all are poisonous. As would be expected, the degree of potency varies among species, but even the least venomous is dangerous to man. Scientists report there is one known species that has a venom at least fifty times as potent as that of the king cobra!

A unique feature of the bite of the sea snake is that it is slow-acting and virtually painless. The lack of reports, especially in beach areas, may often be due to the victim's failing to realize he has been bitten. The delay in receiving medical attention is usually fatal.

Many Asians value sea snakes for their skin and meat, and fishermen take them by the hundreds of thousands each year, working their nets with bare hands. Here is where the greatest number of fatalities occur as many are bitten and few receive immediate medical attention. Still the fishermen refuse to change their tactics.

Since the annual harvest of sea snakes is so great, it is natural to assume that this would affect the sea snake population. Although scientists cannot accurately estimate their number, there seems to have been no population decline. To illustrate their abundance, in 1932 ships sailing in the Strait of Malacca observed a continuous aggregation of sea snakes covering an area about ten feet wide and over seventy miles long—there must have been millions of them. All were swimming determinedly in one direction toward an unknown destination, probably their breeding grounds. Recently the number of reports from ships that have sailed through seas literally alive with sea snakes has increased.

• •

A Tale of Legends

Johnny Weissmuller's original role in *Tarzan, the Ape Man*, released in 1932, involved a safari searching for the legendary elephants' graveyard, which, of course, they found. This subject has been featured in many novels as well as films, and there are probably still people who are searching for the place where elephants, feeling death approaching, go to die. The amount of ivory accumulated there through the ages would be fantastic. The one major flaw in this legend is that the elephants' graveyard simply does not exist!

The legend undoubtedly arose from the relative rarity of elephant remains found in the wilds, for obviously such massive remains would be a long time in breaking down and could not go unnoticed for long. But the scarcity of elephant remains can be explained. Old elephants often go off into inaccessible jungle to die. At times they are escorted by two other big

bulls, which stand on either side of the old one, letting it lean on them. Arriving at a satisfactory destination, usually where food and water are close at hand, the young bulls leave the old one to its fate. It will stay there until it becomes too feeble to feed and then will quietly languish and die. Scavengers quickly reduce the carcass to a pile of bones. During the rainy season floodwaters will deposit tons of silt over the skeleton, and the chances are excellent that the elephant's bones will probably never be located.

When younger elephants are sick or wounded and feel death hovering, they seek out water to soothe their feverish bodies. The flesh of those that die is then eaten by crocodiles and fish. In time, river or swamp silt will cover the skeleton, tusks and all.

Many native ivory prospectors, aware of this practice, make an adequate living by prospecting along the shores of old and new rivers as well as old water holes. Through the years an enormous amount of ivory has been recovered this way. In selling their recovered ivory to the "civilized" traders, their compensation is often somewhat below fair market value. Poaching, the ivory prospectors have discovered, is more profitable.

Scientists have observed that this tendency of elephants to seek out a water hole during illness is rather ancient. Paleontologists in search of fossil elephant material, whether it be mastodon or mammoth, prospect the shores of ancient lakes. A scientist from the University of Arizona was once able to locate the remains of four mastodons in the sediments deposited in the shoreline area of an extinct 500,000-year-old lake. Doubtless any mastodon feeling ill and feverish would seek the cool comfort of the waters of this lake. Those that did not recover are still there. Since this was during the middle of the last Ice Age, there were no crocodiles among the scavengers. Although fish undoubtedly feasted on the flesh, very little disturbed the bones, and they were soon buried in the ancient mud.

All this does not, however, prove that there is no such thing as an animal graveyard. Observations of animal remains indicate that possibility, as reported in the following accounts.

Surgeon-Commander G. Murray Levick, while serving as physician with an Antarctic expedition, was temporarily separated from his forces at Hell's Gate near the Drygalski ice barrier. To his amazement he found there what appeared to be a seal cemetery. On a large patch of ground in front of the doctor and his companion a great number of seals were lying together, frozen solid and in many cases mummified. An examination of the bodies led to the physician's verdict that the seals' remains had been accumulating here for centuries. The throng of dead seals, of varying ages and degrees of mummification, suggested that many generations of seals had crawled from the sea to this particular lonely spot to die.

Some years later Dr. Robert Murphy of the American Museum of Natural History came upon another strange find on the island of South Georgia in the drift ice region of Antarctica. At the crest of a hill he discovered a small, clear lake of melted snow. Around it several penguins were standing, all quiet, unresponsive, droopy, and apparently fatigued beyond revival. Murphy walked to the edge of the lake and looked into it. The incredible spectacle that he beheld, in the depths of the water, was hundreds, or even thousands, of dead penguins, lying prone with their flippers outstretched. Unfortunately he didn't wait to record the final moments of the haggard penguins standing around the lake. So this may be added to the profusion of penguin fact and lore that increases our wonder and admiration of their life—and death—style.

• •

The Fall and Rise of the Black-Footed Ferret

In 1978 the last black-footed ferret, one of six that had been "rescued" in a hapless captive breeding program, was pronounced dead, and the species, along with the recovery program, was considered extinct. Therefore, in 1981, when a black-footed ferret scurried across a ranch in Meeteetse, Wyoming, hope was rekindled and the species was reborn. The fact that Shep, the rancher's dog, had killed the ferret was an unfortunate detail, but it couldn't obscure the larger fact that ferrets were living and thriving somewhere in the neighborhood.

In the years that followed, research teams monitored the population, and by 1984 they had identified 128 ferrets. They were living well and roaming freely in prairie dog burrows (their natural habitat), having first feasted on the occupants (prairie dogs being their almost exclusive prey).

Evidently the species was quite successfully making an ecological comeback, with minimal meddling from its good friend man. The reprieve was exciting but all too brief, for by the end of 1985 the population suddenly dwindled to about twenty ferrets. Just what had happened to cut short this promising return of the species?

The ferrets' undoing was apparently a result of their diet and lifestyle. Being dependent on prairie dogs for both food and housing, the ferrets could thrive only as long as the prairie dog prospered. Unknown to the ferrets, this kind of overspecialization is a principal cause of extinction. As if to prove that point, nature selected that season to deal a double whammy to the ferret with back-to-back epidemics. First it was the flea-borne plague (in humans, the bubonic plague) that decimated the ranks of the prairie dog towns and reduced the ferret population by 50 percent. As soon as the plague was controlled, along came an epidemic of canine distemper that almost wiped out the remaining ferrets.

Now was the time for man to come to their rescue, so six of the remaining ferrets were taken into protective custody by the Fish and Wildlife Service. One, unfortunately, had contracted distemper and passed it on to the others, since they weren't in quarantine. All six died, confirming the fact that canine distemper is almost 100 percent fatal to the ferret.

Another six ferrets were collected, quarantined, and vaccinated, and by March 1987 there was a total of eighteen, of which eleven were female, mostly of breeding age. Happily, 1988 was a baby boom year for captive ferrets, and the population grew to fifty-five after two breeding seasons.

To reintroduce ferrets into the wild a total of 200 to 250 breeding animals is recommended. Another requisite is one healthy prairie dog town covering a total of 7,000 to 8,000 acres for each fifty ferrets. Five of these settings, located judiciously in the several states where ferrets flourish, would help to guarantee a permanent return.

A most welcome bonus to the prospering company of ferrets would be the discovery of other specimens in the wild. This is not a phantom hope, for several have been sighted by responsible but restrained witnesses in areas hospitable to the species. The contribution of a wild bunch to the limited genetic pool shared by all of the captive-bred ferrets would be the best assurance of their success. There is no doubt that the black-footed ferret is on the rise, that the experts have learned much since the early days of ferret bungling, and that ranchers may again look forward to the sight of other black-footed ferrets scurrying across their land.

• •

CHAPTER FIVE

"The Long Island Express"

Usually the great Atlantic hurricanes are not felt severely in the northeastern United States. In fact, severe hurricanes rarely strike the Atlantic coast north of New York City. Tropical storms either move inland long before reaching New England or swing back into the Atlantic. But few people will forget the great storm of 1938.

On September 21, when the storm first struck New York, it was not technically a hurricane, for its winds registered only sixty-five miles per hour. Then, when it moved on to Long Island, the winds quickly surged to 110 miles per hour, and the storm was soon nicknamed "The Long Island Express." A short time later the Blue Hill Observatory near Boston recorded winds up to 180 miles per hour!

Shore areas were hit hard, with Providence and Narragansett Bay serving as central targets for the hurricane. At Providence the water climbed thirteen feet above predicted levels and coursed through the business district at depths of over eight feet. The waters rose very rapidly; in one instance a man crossed the street in ankle-deep water to rescue a boy stranded on top of an automobile. Upon his return, which was no more than two minutes later, the water had reached his chest!

The wind continued to whip the land at a velocity of over 200 miles per hour. It carried saltwater so far inland that, in an area of Vermont 120 miles from the sea, windows were whitened by oceanic salt.

Acres of trees were uprooted and, along with telephone poles, lay on the ground like scattered matchsticks. Automobiles were destroyed by the thousands. Houses were flattened as though crushed by steamrollers. People were killed outright by pieces of glass, bushes, and sticks, all of which flew through the air like bullets. One war veteran remarked, "It sounded like I was back in combat." In a way he was correct, but this time the enemy was nature.

When nature's wrath was spent, the toll was about $350 million in damage and destruction, including 26,000 autos, 20,000 miles of electric wire, and over 200 demolished homes. Nearly three million trees were uprooted or splintered, and beyond all that, the storm took over 680 human lives.

It was the first severe hurricane to strike New England in modern times.

Had the people appreciated its peril, many of the deaths could have been prevented. Since then a much improved warning and protective system has been set up. There have been at least four major storms since 1938, but none has extracted such a toll.

With a perverse kind of humor nature finds time to jest and play a few pranks even in the face of such destruction. In one instance a two-story house was blown end over end for a distance of half a mile. When it finally came to rest, upside down, not a single window pane had been broken. In the same storm the sea swept away a house containing a man's wooden legs. Both were found a week later, twenty miles away, lying on the sand together and still usable!

One survivor returned to his homesite and found his house about 200 yards from his lot. It was sitting on the foundation vacated by another house, but facing in the opposite direction. Inside the house all lamps were still on the tables, and a water jug, which he had placed on the kitchen sink, had not spilled.

A homeowner from Westhampton Beach has become a legendary figure of the 1938 hurricane, and people still talk about him. It seems that when his mail arrived early in the morning of September 21, he received the brand-new barometer he had ordered from a major New York sporting goods store. Unwrapping the package, he marveled at the instrument but found it registered "hurricane." He was outraged at the idea of his new latest-model barometer already malfunctioning. He tapped and shook the barometer, trying to get it to work properly, but the needle remained "stuck" on hurricane. Thoroughly provoked, he wrote a most disagreeable letter to the sporting goods store, rewrapped the barometer, and returned it along with the letter. He then marched straight to the post office and mailed it. To add to his bewilderment, when he returned home, his house was gone!

The most ruthless force of destruction that occurred during this great hurricane was not due to nature but was rather the work of humans. For after the storm had subsided, along came the looters. A witness, marooned on the third story of an office building in Providence, observed from the window the working of the plundering mob. He watched as the looters arrived: swimming, wading, in boats, rising out of the water, they entered through demolished store windows. At first he saw only a few, but soon they came in hordes. They were brazen and insatiable; swarming like rats, they took everything they could lay their hands on. A few policemen came by in rowboats looking for people who needed help. The looters completely ignored the officers. Knowing they outnumbered the police, they just continued with their work of destruction. When they had finished, not a single item of consequence remained in any of the markets. Man was again the final destroyer.

• •

Suzie

For at least seventy million years large marine animals have sometimes deliberately committed suicide by beaching themselves. This bewildering practice continues today and is still an unresolved mystery of the sea.

In Nye County, Nevada, near the town of Berlin, is the Ichthyosaur State Monument. Here visitors can view the remains of six marine reptiles that died at this location about seventy million years ago.

At that time much of North America was covered by warm, shallow seas, with a shore and dry land existing in what is today central Nevada. In these seas swam the ichthyosaur, a large marine reptile whose life was spent in the ocean. Its body had evolved and adapted so thoroughly to a marine environment that, like the whale of today, it could not exist on land and, like its modern whale counterpart, had to come to the surface to breathe.

The six specimens were about the same size as pilot whales and apparently also shared with the whales the unfortunate habit of beaching themselves. The massive size of their chests, unsupported by water, hindered breathing and they slowly suffocated. In fact, visitors to the monument can observe that the nose of one ichthyosaur skull is partially buried downward, showing its state of agony as it tried to breathe. The gentle lapping of the waves pushed the carcasses up on the beach and oriented them with the beach line much as logs always lie parallel to the water's edge. As the bodies decayed, the bones were deposited in the soft ooze and eventually were covered with deposits of alluvial material.

Seventy million years later, on December 31, 1978, a group of fifty-six sperm whales swam ashore and beached themselves on a remote coastal area of Baja California. Unable to support themselves on land, their great weight soon crushed them to death. The largest recorded stranding of sperm whales occurred in 1974, when seventy-two giants swam ashore in New Zealand.

Many scientists now believe that whales often beach themselves when they feel death approaching, either as a result of illness or when trying to escape an enemy. Such a theory was partially proved by a young female pilot whale named Suzie. Pilot whales in particular are known for their tendency to beach themselves. Their skin is exceptionally sensitive to the sun and quickly blisters when they are out of water. Unable to dissipate body heat effectively, beached pilot whales die rather rapidly.

On June 13, 1973, when nine pilot whales beached themselves near Marathon, Florida, only one was still alive by the time the Florida Marine Patrol arrived at the scene. It was a young female that seemed determined to die and vigorously resisted all attempts to tow her out to sea.

Eventually the whale was taken to the Flipper Sea School, where a veterinarian treated her continuously with medications to ease her sunburned skin and fed her food highly laden with antibiotics to prevent infection. Responding to the treatment, she was christened Suzie and seemed to be well on her way to recovery.

For the next few weeks Suzie seemed to be enjoying life. She played actively in the large pool at the sea school and fed quite well. But forty-two days after her rescue she began to refuse to eat her specially medicated food, and nothing could be done to induce her to take nourishment. Three days later Suzie was dead.

An autopsy showed that she had died of pneumonia. It was noted, and is perhaps of some significance, that Suzie's body was found in the shallowest end of the pool. Many of the scientists present believe that, as she felt death approaching, Suzie was again trying to beach herself!

• •

Watch for Dead Branches

Scientists spend much time studying the aerial acrobatics of some primates, those four-limbed creatures that travel so effortlessly through the highest trees. Among the most outstanding of the tree-dwelling trapeze artists is the gibbon. The ease and speed with which it moves through the trees is nothing short of extraordinary. However, even this most efficient acrobat can misjudge distances or at times the strength of a branch as it leaps from tree to tree. The possibility of falling and being injured is always present, because this is the gibbon's mode of living.

A number of scientists currently studying the gibbon's lifestyle have found that aerial accidents are not uncommon. This may seem strange to the casual observer, for gibbons appear to be unerringly accurate as they leap from branch to branch, often at a greater speed than a person can run. Moreover, the flight of the gibbons is often depicted in documentary films.

Apparently gibbons have never seen these documentaries, for recent research definitely indicates that one out of four adult gibbons has broken at least one bone during its lifetime. A number of them have experienced frequent accidents, and among the individuals studied, several showed as many as seven healed fractures. Since these were the specimens that had lived to "talk" about their accidents, one can only wonder how many such aerial accidents are fatal.

• •

Overkill!

Many scientists believe that the extinctions of many Ice Age animal species were due, to a great degree, to man's irreverence toward life and his predisposition to kill far more than he needed for food and clothing. Many kill sites in Eurasia and North America appear to support the premise that wanton destruction was easier than selective killing.

Mass slaughter of horses was an almost continuous event 15,000 years ago near Solutre, France. At the base of a high cliff incredible numbers of crushed horse bones have been found. Authorities estimate that approximately 100,000 horses met their deaths at that locale. Time and time again Cro-Magnon hunters, probably with the aid of fire, stampeded throngs of wild horses off the cliff's edge. Those that survived the fall were no doubt dispatched by hunters who had been stationed below to complete the task.

This is an extreme example of early man's propensity for wholesale butchery, but it is by no means the only one. The amount of food and clothing material obtained from such carnage must have been minuscule compared to the number of animals that were slaughtered. Scientists refer to these events as *overkill.*

Overkill is a technique that seems to have been used by man the hunter wherever he settled, as was clearly indicated in the Plano kill site discovered during the 1950s.

Colorado experienced an intense drought during the 1950s. The wind blowing over the dry soil in a field near the town of Kit Carson laid bare most dramatic evidence of a hunt that had taken place about 10,000 years ago. It is now known as the Plano kill site. Here a group of hunters drove a herd of bison into a ravine twelve feet wide and eight feet deep, killing no fewer than 193 of the herd. The evidence uncovered during the excavation was so rich that the scientists easily reconstructed the hunt in glaring detail.

On that day many thousands of years ago a group of Paleo-Indians surrounded a herd of bison and, screaming like banshees and waving spears and skins, drove the animals into a panic. The bison immediately closed ranks and stampeded toward the long, open ravine that lay to the south, the only direction away from the hunters. They surged into the dry ravine, which was deep and broad enough to trap a good many of the frightened animals. Falling on top of each other, they completely filled the ravine, which then became a flailing mass of animal flesh. The bodies of the trapped bison served as a bridge over which a fortunate few from the last of the herd crossed and escaped.

Ten thousand years later the story was brought to light by concentrated excavation of the kill site. Based on the fact that a significant number of bones were from very young calves, which are usually born at that time of the year, the dig revealed that the hunt must have taken place in late May or early June.

Numerous stone points found among skeletons at the easternmost end of the death trap indicated that a number of skin-waving spearmen had been stationed to the north and west. Their noisemaking activity kept the bison from wheeling around and escaping. The ambush was indeed cleverly planned.

The hunters must have had a field day as they rushed up to the mass of struggling, helpless beasts and exercised their spearing arms. The bison on the bottom of the pile never felt the pain of a spear entering their body, as they were simply crushed to death by the weight of those atop them.

The butchering and skinning that followed must have continued for days. The scientists who worked the site estimated that no less than 60,000 pounds of meat must have been butchered there. Considering that there were probably no more than one hundred people in the entire tribe, this is an incredible amount of meat. Surely much of it had to go to waste, and many of the carcasses on the bottom of the pile were never even touched by the hunters!

This is a dismal but excellent example of overkill, which remained almost unchanged for thousands of years. It didn't take the American Indian long to realize that the hunt could be more productive with the jump method, such as the stampeding of bison over a cliff or into a deep gully. There was certainly plenty of prey, and game killing was perfected to a truly grand operation. This method, along with surrounding the game, was to persist among Plains Indians into historic times. Such a hunt was witnessed and described by Lewis and Clark in 1805.

The Plains Indians, mounted on horseback, were doubtless quite efficient in stampeding herds over cliffs, for several sites have been located. Charles M. Russell, the artist who re-created life as he observed it in the pioneer West of the 1880s and 1890s, was the creator of the painting *The Buffalo Drive*. It illustrates a tense moment as a couple of braves, mounted on fast ponies, wave skins and drive a small herd of buffalo over a cliff. Two Indian women are crouched at the edge of the cliff in full view of, but away from, the action. They too appear tense in response to the exploits they are witnessing and possibly in anticipation of the hard labor and feasting to follow.

• •

Zealous Sharks

Victims of a shark attack usually suffer injuries to the arms or legs, probably because flailing appendages attract the shark's attention. But not always. The following is a true account of a man who was almost decapitated by a shark and lived to tell about it.

A native of Thursday Island, whose name was Treacle, was a professional pearl diver working in the waters of the Torres Strait somewhere between New Guinea and Australia. One day during the year 1913 Treacle was diving

toward a promising-looking oyster when he literally swam right into the wide-open mouth of a huge tiger shark.

The shark seemed to come from nowhere and with jaws swung apart promptly seized Treacle's neck and shoulders, his head completely inside its mouth. To the shark this was a predator's dream; rarely can a man-eater simply open its mouth and have an accommodating meal swim right in.

Treacle's fate was not, however, quite sealed. In desperation, acting on his instinct for self-preservation, he rammed his thumbs into both eyes of the shark. Badly hurt and temporarily blinded, the shark released its prey and swam away. Treacle floated to the surface with his head almost ripped from his body.

When he was found by rescuers, Treacle's shoulders and neck were badly shredded by the slashing teeth. His jugular vein lay completely exposed in places, but by some trick of fate the vein itself had not been severed.

In the hospital Treacle's recovery was touch and go, but he did live, and from then on he exhibited his impressive scars to tourists who flocked to him to see, hear, and believe his story. Treacle's story became a ritual chant for him, and he made a much more prosperous living this way than he had ever made as a pearl diver.

Some Australian great white sharks are more determined than Treacle's tiger shark. They are not so willing to give up a prospective morsel. A thirteen-year-old boy surfing just off an Australian beach suddenly felt something seize his right leg. He kicked at the thing, which, he stated later, felt as though it was attached to his leg. The attachment was an eight-foot white shark. The boy kicked and punched and, finally, in desperation, leaned over and bit the shark on the nose. Still it would not let go.

Lifeguards arrived on the scene and clubbed at the shark with surfboards. The fish must have considered the boy's leg a rare delicacy, because it doggedly held on.

Finally the boy was carried onto the beach with the shark still attached to his leg. It was only after the shark had been clubbed to death that the lifeguards were able to pry its jaws loose. The boy survived his ordeal, fortunately with his leg intact. And just as Treacle gave up pearl diving, the boy gave up surfing.

About a month later on a nearby beach there occurred another unusual encounter with a very determined great white shark. A woman in water about four feet deep was horrified to see a large shark fin cut the water and swiftly bear down on her. After making it safely to the beach, the terrified woman continued running. The shark, excited by its prospective prey, followed her out of the water and onto the beach! There it thrashed around in an almost helpless state until it was finally killed by a lifeguard with a sledgehammer.

• •

Rip Van Winkles

Specific environmental conditions are necessary for a plant seed to germinate; otherwise it will remain dormant. Just how long a seed can remain inactive has been the subject of much scientific research. Apparently the generative capacity of some seeds can be preserved in a dormant state much longer than might be suspected.

In 1967 scientists discovered the seeds of an Arctic tundra bush in a frozen lemming burrow. After applying radiometric dating procedures on the seeds, they discovered that the seeds had been lying in the burrow since the last Ice Age, about 10,000 years ago. Yet when scientists placed them in conditions favorable for growth, the seeds began to germinate within forty-eight hours!

Another example of seeds with suppressed generative capability was discovered recently in Minnesota. Bacterial spores were embedded in muds lining Elk Lake in present-day Minnesota. Samples of the mud were collected in 1983 by scientists from the United States Geological Survey and taken to their laboratory in Denver for research. Radiometric dating of the muds placed the spores at about 7,500 years of age. At the lab the spores were separated from the mud and warmed from their frozen temperatures to surface air temperature.

The thawed-out spores were then placed in a nutrient-rich culture, where, to quote the scientists, "they grew like crazy." One of the scientists involved with this intricate bit of research offered a theory to explain, at least partially, the survival of such ancient bacteria. It was suggested that organisms such as these can survive because, when necessary, they can lower their rates of metabolism, thereby sustaining themselves until conditions again are conducive to growth.

• •

"The Squirrels Are Coming"

Mass migrations of animal species such as the lemming are well known. Sudden mass migrations have also been known to occur among other animal groups, such as the gray squirrel.

In the early days of our country's history mysterious mass movements of gray squirrels occurred periodically across the United States. For no apparent reason hundreds of thousands of the squirrels would band together and move like a flood of gray fur across the countryside, devouring all the nuts and grain that lay in their path.

In 1749 a horde of migratory squirrels caused so much damage in the cornfields of Pennsylvania that authorities offered a bounty of three pence for each squirrel killed. As a result well over 640,000 squirrels were destroyed by money-hungry hunters. Even this slaughter made scarcely a dent in the multitudes, and hundreds of thousands survived, continuing their march over mountains and plains, clearing fields and forests.

No records were kept of the final destination of the squirrels, and nobody seems to know what happened to them. They simply "vanished."

• •

When a Day Was Short

Many scientists agree that during the primal stages of the earth the moon was much closer to this planet than it is at present, possibly as close as 12,000 miles. At that period of earth history a lunar month, the time it takes for the moon to revolve around the earth, would have been as short as six-and-a-half hours.

The gravitational effects that the earth and moon had on each other must have been tremendous. They resulted in a braking effect, and with the

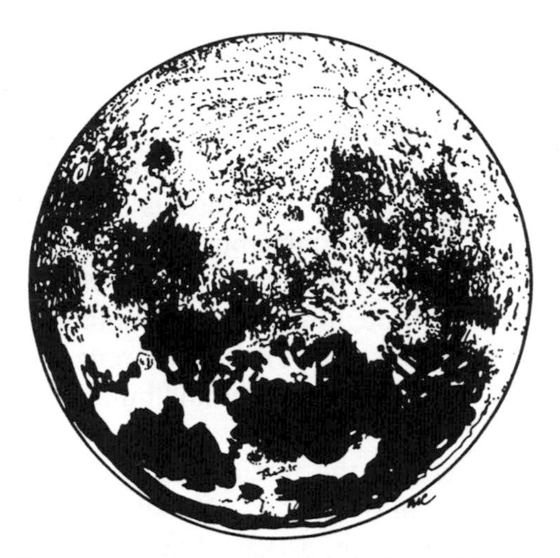

passage of time the axial rotation of each celestial body was slowed down considerably. The centrifugal force of the earth resulted in pushing the moon farther and farther away from the earth. Even now, over four billion years later, it is still receding at the rate of about four inches per month.

Strangely, the evidence for this phenomenon is provided by living and fossil organisms. Many modern coral species show well-defined growth bands. Between major growth bands are very fine bands that always number 365. Obviously one fine band of growth occurs every day of the year; the coral adds to its skeleton on a daily basis.

Certain fossil coral species show the same characteristic and thus record in their skeletons their prehistoric rate of growth. However, they contain many more daily growth rings than modern species, indicating that there must have been a greater number of days in a year than there are at present. Such evidence suggests that the earth in prehistoric times must have rotated more rapidly than the current rate of once in twenty-four hours.

A study of corals of 370 million years ago shows the same growth ring characteristic. Since they contain 400 fine daily bands, the implication is that at that time a year was 400 days long!

The older the fossil is geologically, the greater the number of growth bands it contains. By computing this rate of growth, many scientists now believe that during the primal stage of the earth, about 4.6 billion years ago, a day was only about four-and-a-half hours long!

• •

The Greatest of Rivers

The Amazon is by far the world's greatest river, navigable by oceangoing ships for over 2,300 miles. Vessels from all nations make regular trips completely across Brazil and into Peru or to any of the countries of the Amazon headwaters. Yet it is a river of very humble origin.

High in the cloud-piercing Peruvian Andes, only about seventy miles from the Pacific Ocean, a tiny trickle of water flows from under an ice field 17,220 feet above sea level. At first the water is only ankle deep, but as it cascades down the mountainside numerous rivulets join in and its volume grows rapidly. Soon it is a full-fledged mountain river, churning muddy brown and red as it roars through numerous mountain gorges. For its first 600 miles the Amazon drops over 27 feet per mile. Finally it squeezes through a gap called the Gateway of Fear and bursts onto the jungle as a great hydraulic explosion.

As the river snakes across the torrid jungle, over 1,000 tributaries, 50,000 miles of streams from six nations, virtually dump water into it. Most of the tributaries from the north, darkened by the decaying foliage of the rainforest vegetation, flow black. Those from the west, carrying the sediment from the Andean highlands, are light chocolate in color. The clear water filtered through the white, sandy country to the south reflects a deep blue. For most of its length the rainbow river exhibits constantly changing hues, the colors being determined by its tributaries as well as by season and even time of day. In general the Amazon is a muddy medium brown.

After its rampage down the mountainside the Amazon becomes more leisurely, descending only three inches per mile as it moves across the broad jungle basin to the Atlantic Ocean. Torrential rains are added to the river, swelling it to such proportions that it resembles a moving inland sea. It drains an area three-fourths the size of the United States, with a rainfall averaging over 80 inches per year. It is scarcely surprising that in volume it exceeds the next eight largest rivers combined.

The total length of the Amazon is roughly 4,000 miles. In places the banks are sixty miles apart, with the main channels often so wide that sailors standing on the decks of their ships are unable to see the distant shoreline. At its mouth the river is 208 miles wide, more than the distance from Boston to New York.

Draining nearly half of South America, the Amazon empties more than three billion gallons of water per day into the Atlantic Ocean. The volume is so tremendous that 150 miles from the mouth of the river fresh water can still be dipped from the ocean and drunk. Fifteenth-century explorers who accompanied Columbus observed that, far from sight of any land (and they presumed the nearest land to be India), they were sailing through fresh water. Incredibly, fresh water from the Amazon is still detectable 200 miles at sea.

Since the original source of most water is the ocean, it would appear that the great river is merely returning to the sea that which was taken from it.

• •

Oldest Known Paddle

The area now known as Yorkshire, England, was once a camping site for Stone Age hunters. About 10,000 years ago a small group of people established a campsite near the shore of a now-extinct lake. That they were very active hunters is indicated by the numerous remains of game animals such as deer and elk. Strangely, although they lived beside a large lake, no fish bones were found in the camp debris. Scientists believe they did fish or at least that they traveled on the lake. This is definitely indicated by the excavation of a wooden paddle. Although relatively insignificant, this artifact is the oldest implement of water navigation to have been discovered anywhere in the world.

• •

Fish Food

The havoc caused by swarming locusts is familiar to people the world over. Some may have witnessed the plague of the Great Plains in the 1930s, compounding the devastation of the dust bowl. Others may have experienced the African locust swarms, ten to fifty billion strong, darkening the

sky and advancing on all growing things like a devouring fire. And who can forget the Eighth Plague over the land of Egypt, recorded with succinct fervor in Exodus?

When innocuous grasshoppers impact their environment they must migrate; at this time they pass into the locust phase of their existence. In their migration stage they reign sovereign over all. With their new expanded size, wingspan, power, and stamina, no thing or creature can stop their airborne invasion. But the gods of the locusts have blatantly overlooked one weapon—the wind. In their mass flights locusts are often at the mercy of the wind and may be driven out to sea where they have no place to light. Mariners have sighted huge swarms of locusts as far as 1,200 miles beyond any land. When their endurance finally gives out, the insects, unable to replenish themselves with food, drop by the millions into the sea.

There is paradoxical retribution in the fact that the locusts, whose visitations have produced famine as they destroyed vast crops, can become, by a malicious trick of the wind, tasty morsels for the creatures of the sea. The despoiler of food does, finally, become the food.

• •

Ancient Cave Tragedies

During the Ice Age when ice was at its greatest expanse, glaciation produced a period of rainfall which vastly changed environmental conditions in the American Southwest. At that time the region became filled with vast swamps and enormous lakes. Today these areas are arid to semiarid, and desert conditions are spreading.

The advancing northern glaciers affected other areas of the country quite differently. In contrast to the watery Southwest, much of Florida was semiarid during the Ice Age.

Modern-day geologic and archaeologic research in Florida employs scuba-diving scientists who explore underwater sinkholes. This type of research is becoming widely pursued, and thus far, the results have been quite productive. Southeast of present-day Sarasota a scuba-diving scientist uncovered evidence of a tragic drama that occurred over 12,000 years ago in a goblet-shaped sinkhole known as Little Salt Spring.

Since climatic conditions in parts of Florida at that time were semiarid, the water level at Little Salt Spring was about ninety feet lower than at present, exposing a large hole in the ground. Into this hole an ancient Indian fell or was pushed. Either a short time before or after the mishap, a land tortoise also became a victim of the yawning aperture in the ground. The water at the bottom absorbed their fall, and they both survived.

The man was able to find shelter on a ledge under a large protruding overhang. He captured the tortoise, cooked it, and lived on the meat. But he lived more on hope than reason, because eventually the meat would all be consumed and the overhang would prevent his climbing out. He could only shout, hope, and wait, and wait, and wait. . . . Eventually he was found—12,000 years later.

Similarly specialists in cave exploration often uncover evidence of strange and unusual events. One such cave expedition uncovered the calamity of a man who disappeared from the company of his contemporaries ages ago. Researchers refer to this event as "The Tragedy of Lost John."

In 1935 a group of speleologists were exploring an unknown corridor in Mammoth Cave National Park, Kentucky. Approximately two miles from the entrance they came upon an Indian mummy that lay partially hidden beneath a six-ton boulder. Several archaeologists were brought in for consultation and were able, with little difficulty, to reconstruct what had happened to "Lost John," as he was appropriately nicknamed.

The victim had worked his way along the dark passages using a reed torch to light his way. It is quite likely that he was searching for gypsum, which was used by his tribe for ceremonial paint. As he crouched on a ledge, his foot must have accidentally dislodged a rock, causing an enormous boulder to crash down on top of him. And so he passed from the sight of mankind—for the next 2,000 years.

• •

Gourmet

Most people who have fished in marine waters have, at one time or another, witnessed the catch of a puffer. These fish appear small and harmless enough, but curiously, when alarmed, especially after being hooked, they inflate themselves to three times their normal size. They are something less than the ideal catch, because many species are quite lethal.

In Japan, however, the puffer, the honorable fugu, is the ultimate in gourmet dining. Despite its being ugly and deadly poisonous, the fugu is eulogized as delicate, subtle, smooth as silk—indescribably delicious!

How poisonous can it be? Have the Japanese truly been risking terminal indigestion for hundreds of years? According to chemical analysis, the fish contains tetrodotoxin, an anesthetic over one hundred times more potent than cocaine and twenty-five times more powerful than curare. The white powder found in one medium-sized fugu, one-tenth the weight of an aspirin, is sufficiently poisonous to kill thirty people. A lethal dose of one milligram could be served on a pinhead, making the fugu without doubt the most deadly creature ever to cross a dinner plate. And there is no known antidote.

One authority, supporting his claim of the puffer's potency, cites an incident in which a man merely picked up the roasted entrails of a puffer and was in a coma for twenty-four hours. A restaurant owner has described the terrible death following consumption of the fish, in which the victim can think clearly but cannot speak or move, and eventually cannot breathe. Such was also the experience of Captain James Cook, who with several of his crew tried the merest taste of the roe, and all "were seized with weakness, numbness, no sense of feeling, vomiting and sweat. The pig that ate the entrails was found dead the next morning."

The secret to surviving this ultimate feast is in its preparation. The edible portion of the fugu is the flesh, provided that it has not been contaminated by the liver, ovaries, intestines, kidneys, skin, or eyes. Qualified fugu chefs must undergo rigorous training, pass difficult examinations, and be apprenticed for two years to be licensed.

Beyond the mere preparation of the fugu to remove all toxic portions is the artistic presentation of the dish itself. Cut very thin and arranged in exquisite patterns on a platter, the fugu appears to justify the price of $150 to $200 for a single dinner. The ultimate disgrace for any chef would be to have his customer carried out of the restaurant feet first. Compared with this, the returning of an improperly prepared dish to the kitchen is relatively trivial.

Despite the care and training that attends the eating of fugu, over 200 gourmet Japanese have died during the past ten years from eating a fugu meal. The record among experienced chefs is excellent, and most of the fugu deaths are the result of improper preparation at home. Others have happened when a devotee of fugu pleads with the chef for the ultimate thrill, such as the detoxified (undependable, with no guarantee) liver or ovaries. This routinely becomes the true "ultimate" meal!

Why then, are so many people willing to risk death? Why is the reservation list so long that people wait months for their chance at a fugu dinner? Is this adventure in good eating also an experience in daring, an eastern version of Russian roulette? Is the flesh that tasty? Or is it really because the fugu, like most risky foods, is thought to be a powerful aphrodisiac?

• •

A Case of Apple Polishing

The practice of "bringing the teacher an apple" did not originate with the modern schoolchild. Actually it is a very old custom that was practiced in ancient times when some long-forgotten student discovered that poor

performance might be ignored in proportion to the favors presented to the teacher.

Ancient Sumer calls up visions of crumbled walls and ruins, but the daily life of the Sumerians is well recorded in the writings left by the people of that long-gone age. It could be said that by etching their clay tablets the people often recorded a form of diary.

Outstanding is a series of tablets left by a schoolboy who was having some difficulties. His earliest account tells how he recited his tablet, ate his lunch, and prepared a new tablet. One can almost envision the young boy shaping and smoothing his tablets of clay and copying his lessons for the day, probably from a master tablet.

As the boy's record continues, he appears to have done something wrong that precipitated a rather bad day at school, because he was repeatedly flogged. In desperation he asked his father to invite the teacher to their home for an evening meal.

The father did just that, and the teacher was wined and dined and even given a new garment. The scheme apparently worked; as he was leaving, the teacher proclaimed that the boy was already "becoming a man of learning." The scientist who translated the boy's tablets refers to this account as the earliest case of apple polishing. And this happened at least 4,000 years ago!

• •

Who's Out There?

It is common knowledge that astronomers measure distances in outer space by light-years, the distance light will travel in one year at the rate of over 186,000 miles per second. Reducing this to terms that make sense to the nonastronomer, one must translate a light-year into a common unit of measurement, such as miles.

As an illustration, consider the distance from the moon to the earth, approximately 240,000 miles. Allow the thickness of an ordinary playing card to represent 240,000 miles. By constantly adding cards to the stack, each representing the distance from the earth to the moon, it would take about nineteen miles of playing cards, at six million cards per mile, to reach the distance from earth to the nearest star, Alpha Centauri, 4.3 light-years away.

The earth is part of the Milky Way galaxy, at some distance southwest of the center. If the stacking of playing cards were to continue until the equivalent distance from earth to the center of the galaxy were reached, a continuous stack could be assembled that would go five times around the equator of the earth (something like 120,000 miles of cards).

The number of stars contained in the entire universe is impossible to comprehend. Our galaxy alone, which is only medium-sized, contains over one hundred billion stars. Multiply this by the ten billion or so other galaxies within observable limits, some of which contain many times that number, and the only star count possible is "beyond our comprehension."

Obviously then, space is not empty, but on closer observation it seems not to be incredibly cluttered. Light from the Large Magellanic Cloud, one of our nearest neighbors in the cluster of about thirty galaxies called the Local Group, takes 170,000 years to reach the earth.

The universe itself, or space, is so vast that, traveling at the speed of light, it would take roughly 30 billion years to cross from one of its borders to the other—and 100,000 just to cross our galaxy. And still the universe continues to expand! Considering the infinite size, number, and variety of stars within our universe, it is logical to assume that there are numerous stars with planetary systems, some of which could contain life as we know it.

No criteria prescribe that a star must be the size of our sun, or of any particular size, mass, brightness, or energy range, to have a planetary system. The size of stars varies greatly; some are gigantic by any standard of comparison and certainly dwarf our sun with its 865,400-mile diameter. Recently scientists discovered a faint blue star that is believed to be the largest yet identified. It is at least ten times hotter than our sun, with a diameter about the size of our solar system.

The star, named R-136a, is 150,000 light-years from earth (or a stack of over 3.6 trillion playing cards). It is in the Large Magellanic Cloud, one of the Milky Way's galactic cluster. Scientists believe that in just one second the star puts out more energy than our sun does in five years!

It is intriguing to imagine the forms of life that could be supported by a star of such size, which radiates such heat and energy. No astronomical instruments have yet been developed that could determine whether R-136a, or any other star, is the center of a system of planets.

• •

Control by Fear

In many primitive cultures of the world the witch doctor still holds sway over people, preying on their superstitious minds with occasional deadly results.

Many Australian aborigines believe an enemy can be killed if a witch doctor points a human bone at him while singing a death chant. When the shaman does this, it is thought, an invisible splinter of bone implants itself in the body of the victim and will eventually kill him. Quite often the victim, aware of the shaman's actions and powerless against them, dies of sheer terror. There is, however, a cure. The victim's only hope is that the shaman will go through the motions of removing the deadly, invisible bone. This is done quite frequently—for a price.

Some primitive African tribes believe that witch doctors can kill their enemies by directing lightning bolts to strike them. Strangely, during the late 1920s in Bechuanaland, a charge of murder by lightning was leveled against a tribal witch doctor.

As a result of an argument with one of his fellow tribesmen, the shaman angrily called on heavenly lightning to destroy his adversary. To the amazement of all present the accursed man was suddenly struck and killed by a lightning bolt. Quite probably the one most amazed by the event was the witch doctor himself. Surprisingly, but at the same time lending credibility to his supernatural powers, he pleaded guilty at his trial. The chief punished him severely; the man was whipped and branded.

Remarkable coincidences often give the witch doctor the appearance of supernatural powers and vision. In 1949 a witch doctor in Salisbury, Southern Rhodesia (now Harare, Zimbabwe), was arrested and sentenced to prison for illegally practicing witchcraft. As he was led away to prison, he screamed a curse that lions would return to the city streets.

Salisbury, a modern city, had not seen lions anywhere near its borders for over fifty years. But just three weeks after the man screamed his curse a pride of lions strolled nonchalantly through the quickly emptied streets! The lions remained in the area for two weeks, raising havoc with domestic

cattle before moving on. No humans were molested, possibly because many people were afraid the lions were not of this world and kept a respectable distance between themselves and the visiting marauders.

It is not known if the punishment for the witch doctor was increased after he continued his illegal practice so effectively.

• •

Kohinoor—"Mountain of Light"

It was once thought that he who owned the Kohinoor diamond owned the world. Authorities believe the gem was found more than 5,000 years ago. Indeed the Kohinoor has the longest recorded history and has been the cause of more bloodshed and intrigue than any other diamond, including the famous Hope diamond.

Its first recorded owner was the rajah of Malwa, who came into possession of it in 1304. The diamond was passed on to each succeeding Mogul ruler until 1739, when the Persians under Nāder Shāh conquered India and seized all of the Mogul's jewels—all, that is, except the Kohinoor.

For two months the Persian conqueror ransacked Delhi in an almost desperate attempt to locate the diamond, but to no avail. Finally a harem girl revealed to him that the Mogul Muhammad Shāh always carried it in his turban. By trickery the Nāder Shāh managed to obtain the turban, and clutching it to his breast, he rushed into his tent, where he feverishly undid the turban's silken reels. There was the diamond. He was taken aback by a brilliance much greater than he had expected and exclaimed, "Koh-i-noor!" meaning in Persian "a mountain of light." The diamond has retained this name ever since.

Not surprisingly, the new owner's almost immediate fate was assassination. The precious jewel was passed on to his son, who hid the diamond when he was about to be deposed. In an attempt to force him to reveal the diamond's location he was tortured horribly, but his secret died with him. Despite his resistance, the gem was eventually found and passed from ruler to ruler, several of whom met untimely ends because of the Kohinoor.

No less imaginative were the hiding places of the Mountain of Light than were the tortures applied to possessors to reveal its whereabouts. One diabolic sultan had a plaster container placed around the head of the owner and, as he filled it with boiling oil, entreated him to divulge its hiding place. He too had to wait until the uncooperative possessor died to conduct a successful search. Brother blinded brother, only to be blinded by a third brother. Then, in 1813, the Kohinoor reappeared in India as the property of Ranjit Singh, the ruler.

After his death the British annexed the Punjab region and confiscated everything of value, not the least of which was the Kohinoor. As British property it was presented to Queen Victoria. Disappointed by its crude cut and its lack of fire and brilliance, she had the stone recut. The recutting reduced the Kohinoor from 186 to 109 carats and, most unfortunately, also reduced the diamond's brilliance instead of enhancing it. Nevertheless, Queen Victoria wore it as a brooch until her death, whereupon it was transferred to the crown of Queen Mary. It was later set in the crown made for the coronation of Elizabeth II.

The trail of intrigue and slaughter caused by the desire to possess the

Kohinoor is apparently over. The great gem now resides peacefully with other crown jewels in the Tower of London.

• •

Rocky Road to Riches

It seems strange that one man could have had the wealth of kings only to die a pauper. In 1886 George Harrison, a South African prospector, found traces of gold in the Witwatersrand Basin. Although it looked very promising, Harrison was at the time so desperate for cash that he sold his claim for a mere $50. In so doing he missed the rare opportunity to become a multimillionaire! The buyer took advantage of Harrison's impulsive move, and the accumulated claims that resulted became the South African Rand, the most productive gold-bearing region in the world. During its first century its forty mines yielded 36,000 tons of gold. It is still a major source of gold, as well as silver and platinum, and today produces over 70 percent of the world's gold supply.

When gold-bearing rocks erode, they crumble and release free gold particles that accumulate in stream channels, usually as very small nuggets—with occasional exceptions.

In 1879 two prospectors, almost at the end of their rope and ready to abandon the search, made a most extraordinary discovery. They found, in Victoria, Australia, a rather exceptional nugget. It weighed nearly 150 pounds, the largest single gold nugget yet found. It was a happy, unexpected discovery for them, so they named the nugget "Welcome Stranger." They promptly sold it for about $50,000, which in those days was considered a significant fortune. On the current market the gold alone would be worth millions, but as a single nugget one of this size would be absolutely priceless. The owners preferred to cash it in.

The adventures of this nugget of nuggets did not end there. In the museum of a leading midwestern university there was on display one of the gold-painted plaster models of the "Welcome Stranger." The printed sign next to it read, "The largest gold nugget in the world." In the description it was clearly identified as a plaster cast.

But during a dead-of-night raid a thief stole the cast, undoubtedly thinking it was real. An uninformed person, indeed, because had it been real gold he scarcely would have been able to budge the nugget, let alone carry it hastily from the premises. But his fortune turned to misfortune, as he dropped the "nugget" at the door and the plaster shattered. The luckless thief was so taken back by this unhappy turn of events that he gave himself up to the authorities.

• •

The Last Meal of Portheus

In the museum of Fort Hays State University, Hays, Kansas, a most unusual fossil is on display. It is that of a large tarponlike fish, *Portheus molossus*. Within the fourteen-foot skeleton is that of another fossil, the six-foot remains of another tarpon, *Gillicus* sp., which had evidently been swallowed whole.

The sequence of events was easily established by the investigative research team. *Portheus*, a vicious predator of the Mesozoic seas, was out looking for prey. Spotting the potential victim, it darted down and quickly swallowed its prey alive. As the saying goes, "it had bitten off more than it could chew," because the victim, the six-foot *Gillicus*, did not die readily. Instead it sold its life dearly by intensely thrashing about inside the predator's stomach. The swallowed victim, being as large as it was, inflicted much damage.

Portheus, after swallowing its prey, swam off contentedly, since the prey had thoroughly appeased its hunger. It must have become aware almost immediately of the excruciating stomachache it was developing, and the pain worsened quickly. The internal thrashing of the *Gillicus* must have ripped the stomach apart along with other internal organs. *Portheus* died promptly and sank slowly to the sea bottom, a victim of its own prey. Ironically the swallowed fish, as if in momentary retribution, lived on after its assassin died. Since there was no way it could get out of the body of *Portheus*, it too died shortly afterward. The body of the large tarpon, along with its fatal dinner, settled into the soft bottom muds of the sea, where they were buried and preserved by natural processes. In time the skeletons were fossilized as the encasing muds were turned into rock.

With the passage of time and the ancient sea long gone, the forces of erosion eventually laid bare the rock, exposing the skeletons of *Portheus* and its last meal. The revealed fossils represent a rather dramatic bit of evidence of an event that happened well over one hundred million years ago in an inland sea that covered the area now known as the state of Kansas.

• •

The Day the Earth Died

Nuée ardente, the fiery cloud, a most terrifying phenomenon of volcanism, is a mass of superheated gases (about 2,000 degrees Fahrenheit), incandescent ash, and other volcanic clastic debris. It shoots down the side of the volcano at hurricane speeds, clinging to the ground, instantly baking anything organic. Inhaling it is like breathing fire. It can sterilize a landscape almost as effectively as an atomic bomb. For one who might survive being roasted alive, the gases in the cloud guarantee death by asphyxiation.

When the fiery cloud came shooting down the side of Mount Pelée and enveloped the city of Saint Pierre and its inhabitants, it seemed as if the earth had died.

Basking in the warm tropical sun lay the small Caribbean island of Martinique, a prized French possession. Roughly forty miles long and sixteen miles wide, it was mantled with junglelike forests. The highest peak of the island was a volcano named Mount Pelée. It had last erupted in 1851, sprinkling the countryside with rich volcanic ash and making the soil fertile for sugar, coffee, and tobacco plantations.

The island boasted the beautiful port of Saint Pierre, a flourishing city and the most important commercial center in the Lesser Antilles. It was often referred to as "The Paris of the West Indies." The city housed a population of about 30,000 people who took much pride in their hometown. The charming appearance of crooked, narrow streets bordered with low tile-roofed houses, all set in a background of tropical vegetation, presented the impression of an alluring travel poster. One main street ran the entire length of the city and was crossed by a number of small streets. The local newspaper *Les Colonies* reported the local news to a purpose. In the storage houses of the city was an ocean of rum in barrels and bottles waiting to be exported. And over all this bustle of healthy activity loomed a sleeping volcano.

Mount Pelée, located about six miles northeast of Saint Pierre, was not particularly impressive in the ranks of the world's volcanoes, being only 4,430 feet high. The mountain was a resort to the residents of the city. Its ravines and forested slopes provided enjoyable vacation and recreational interludes, as did its crater lake. A short, pleasant climb through a gap in the caldera led to the lake. Neither residents nor tourists were alarmed about spending their leisure time in the cavity of a volcano.

In the spring of 1901 a group of picnickers, upon arriving at the summit, found a small jet of steam rising from the margin of the crater lake. It smelled of sulfur and had killed nearby vegetation. Nothing more was observed until almost a year later, when, on April 2, 1902, the nightmare began.

On that date a local scientist from Saint-Pierre Lycée, Professor Landes, noticed new fumaroles appearing in the upper valley of a river that ran down from the summit of the volcano, and on April 25 Mount Pelée really awakened from its deep sleep. The people of Saint Pierre were enthralled by the display put on by the volcano as it threw vast clouds of steam charged with ash and rock fragments straight upward from its summit amid thunderous roars.

Ash began to fall regularly on the city and in a very few days Saint Pierre began to take on the appearance of a winter scene in a New England city.

Mild, perceptible earthquakes were common, and people sat down to dinner amid rattling dishes. Air became difficult to breathe as ash continued to fall. Very few people cared to venture forth from the security of their homes. Saint Pierre was becoming a silent city as ash deadened the sounds of footsteps and the rolling wheels of carriages. The first casualties were two horses pulling a large wagon; weak from breathing sulfurous fumes and ash, they fell almost simultaneously in the street, dead of asphyxiation. The eruptions became more frequent and violent, and sleep was uneasy; people were awakened by explosions that shot thick, dark clouds into the atmosphere, dramatically laced with brilliant crisscrossed lightning. Still the residents did not evacuate the city.

Early in the afternoon of May 5 a torrent of boiling mud surged downward from the summit at express-train speed, overwhelming a sugar mill and burying at least forty men alive. The smokestacks protruding above the mud became their monument. Mount Pelée had claimed its first human victims. This was the straw that broke the backs of the people, and citizens began to panic and pack their possessions. Many did leave for the safety of the capital city of Fort-de-France and other points to the south of the island. These refugees were to be the survivors of the forthcoming disaster. Many people who lived in outlying areas fled into Saint Pierre for refuge, a fatal mistake.

An important election was scheduled for May 10, and it was imperative to the island politicians that the people remain in Saint Pierre so they could vote. So the government officials took steps to stop the exodus from the doomed city. An official commission appointed by the island governor reported that there was no immediate danger and that the people were perfectly safe in Saint Pierre; there was no reason to leave the city. As a gesture of assurance the governor and his wife came to Saint Pierre; they never left.

The eruptions continued and became more and more violent, and again people began to panic and tried to leave. On the roads to the south the governor stationed soldiers, who in many cases used violence to stem the flow of refugees from Saint Pierre. On May 6 it seemed that Armageddon was upon them as dark, thunderous eruptions issued continuous black, lightning-laced clouds and the city was in almost perpetual darkness.

What more could a volcano do to announce its intentions? The prayers for salvation seemed to be answered as news came of the eruption of La Soufrière on St. Vincent, a neighboring island to the south. Perhaps this outburst would drain away the hot gases from under Mount Pelée and cause its eruptive state to subside.

The next morning it appeared to the suffering populace that the La Soufrière eruption had done just that, because the dark clouds of ash were gone. The plume that rose from the crater of Mount Pelée was clear, white steam. It seemed that the worst was over and, just as they had been assured, the people were safe. The respite was short-lived.

Mount Pelée quickly came to life again with loud detonations, clouds shot with lightning, and incandescent material pushed up from the crater. Again talk of mass evacuation prevailed, with soldiers actively preventing an exodus. The editor of *Les Colonies*, politically ambitious and aligned with several candidates in the forthcoming election, was energetic and persuasive in reassuring the people. While the paper was reporting the hysteria and describing scenes of destruction, it still declared, "We confess that we cannot understand the panic. Where would one be better off than at Saint Pierre?" The paper quoted, or actually concluded from an interview with the well-respected Professor Landes, that "The Montagne Pelée presents no more danger to the inhabitants of Saint Pierre than does Vesuvius to those of Naples." It was hoped that these messages in the May 7 edition of *Les Colonies*, the *final* edition, would quiet the doubts of the most timid.

It is interesting to note that a sea captain, Marino Leboffe, who had seen Vesuvius in eruption, was anxious to get out of port without delay. On the afternoon of May 7, even though his ship *Orsolina* was only half-loaded with scheduled cargo, and over the objections of customs officials, he hoisted anchor and sailed from the scene of impending chaos to the safety of the open sea.

The wife of the American consul at Saint Pierre wrote to her sister in Boston that day, telling her there was no cause for alarm. She further assured her sister that they would leave if there was a particle of danger and that there was an American schooner in the harbor for just that purpose. She never lived to regret her decision.

Very early in the morning of May 8 Captain G. T. Muggah of the British steamship *Roraima* steamed into the harbor at Saint Pierre. The deck was coated with ash that had landed on the ship while it was still at sea. Standing on the bridge, the captain eyed the volcano suspiciously; only a thick column of steam was issuing from Mount Pelée at that time. The captain turned to a passenger and told her he would not remain in this harbor an hour more than necessary. When the ship did leave, the good captain was not on board.

In the post office of Saint Pierre the night-shift telegraph operator was ending his transmission with the latest official report on the volcano. The telegrapher in Fort-de-France, Martinique's capital city twelve miles to the southeast, was getting ready to begin his reply. The time on the wall clock read 7:52 A.M. "*Allez,*" clicked the operator in Saint Pierre, the signal to proceed. The telegrapher in the capital city pressed his key, but the line was dead. At that very moment Saint Pierre had died.

May 8 was Ascension Day, and the morning was clear and sunny. The air vibrated with church bells awakening the populace of Saint Pierre to this holy day. The people were weary since most had spent an almost sleepless night watching the volatile fireworks display of the angry volcano. At 6:30 A.M. the volcano appeared to be resting because only a column of vapor was seen rising from its crater. A gentle westerly breeze deflected the ash away from the city for the first time in days, and the air was less noxious to breathe. In the harbor, at anchor, lay eighteen ships.

It must have been about 7:50 A.M. when the cataclysm started; four staccato reports were heard first by the sailors aboard the ships. A minute later the volcano blew apart. The clock at Hôpital Militaire stopped forever at precisely 7:52, marking the exact time of the incineration of the city and its 30,000 inhabitants.

The top of the mountain was plugged solidly with a semisolid mass of viscous lava, so the final eruption was on the side of the volcano facing Saint Pierre. Under tremendous gas pressures a fissure on the flank was forced open, and the fiery cloud was shot at the city with hurricane speed. The initial speed had to be well over one hundred miles per hour, and this increased as it raced down the mountainside. The fiery cloud seemed to clutch the ground as it fell forward, its front tumbling over and over, glowing incandescently with lightninglike flashes within its depths. In less than a minute the cloud had reached Saint Pierre, enveloping it like a luminous black sooty blanket and blotting out everything.

Within the nuée ardente itself there is little or no free oxygen necessary for combustion. Thus when it first enveloped Saint Pierre, the city was scorched rather than set ablaze. But when the front of the cloud passed and oxygen returned, a matter of a few seconds, the city took fire. It was then that the incandescent particles set the superheated Saint Pierre to burning. Thousands of barrels of rum exploded with a roar, adding to the chaos, and the flaming streams of rum resembled lava flows. A number of people, witnesses to the initial explosion, rushed to the docks seeking safety. As the superheated cloud passed over them, all became human torches. The city was now completely ablaze from one end to the other.

The ships offshore were anchored broadside to the onrushing cloud and received the fiery impact full force. Most of them capsized and sank immediately, and the hulls of all were instantly afire. Only two ships remained afloat, the *Roraima* and the *Roddam*. The *Roraima* lost its captain, along with its funnel and lifeboats; fire was all over the deck. The *Roddam* heeled until water poured over the rail. She surely would have capsized, but the anchor chain broke, and the ship slowly righted itself. It was on fire from one end to the other.

Assistant Purser Thompson was one of the sixty-eight crew members on board the *Roraima* who watched the magnificent spectacle of Pelée until the explosion. The following is from his written account of the events of that fateful day:

> I saw St. Pierre destroyed. The city was blotted out by one great flash of fire. Of eighteen vessels lying in the road, only the British steamship *Roddam* escaped and she lost more than half of those on board. It was a dying crew that took her out. . . . There was a tremendous explosion about 7:45. The mountain was blown to pieces. There was no warning. The side of the volcano was ripped out and there was hurled straight towards us a solid wall of flame. It sounded like a thousand cannon.
>
> The wave of fire was on us and over us like a flash of lightning. It was like a hurricane of fire. I saw it strike the cable steamship *Grappler* broadside on and capsize her. From end to end she burst into flames and then sank. The fire rolled in mass straight down upon St. Pierre. The town vanished before our eyes.
>
> Wherever the mass of fire struck, the sea boiled and sent up vast columns of steam. The sea was torn into huge whirlpools which swirled under the *Roraima* and pulled her down on her beam end with the suction. The fire wave swept off the masts and smokestacks as if cut by a knife.
>
> Captain Muggah was overcome by the flames. He fell unconscious from the bridge and overboard. The blast of fire from the volcano lasted only a few minutes. It shrivelled and set fire to everything it touched. . . . The blazing rum set fire to the *Roraima* several times.
>
> Before the volcano burst, the landings of St. Pierre were covered with people. After the explosion, not one living soul was seen on the land. Only twenty-five of those on board were left after the first blast.

The full horror of the catastrophe remained unknown to the rest of Martinique for hours. Officials in Fort-de-France were completely stupefied, since no word whatsoever had come from the governor, who was still in Saint Pierre. Every attempt to communicate with the city proved futile. Finally the acting governor sent a warship, which arrived at the burning city at about 12:30 P.M. The expedition on board clung to every theory that would invite hope, but all hope vanished when the captain first viewed the city through binoculars. Although they had feared the worst, they were totally unprepared for the scene of devastation, the appalling silence, the view of "a world beyond the grave."

Shore parties abandoned their efforts to penetrate the city because the heat was so suffocating and the ground under them was like a glowing brazier. They could see, however, that the destruction of Saint Pierre was complete. What had once been "the Paris of the West Indies" now had the appearance of ancient archaeological ruins. They could find not a single building still standing, and only the lower portions of stone walls remained intact. The city was one big heap of dirty rubble, and everywhere one looked there were human corpses—in fact, about 30,000 of them.

Death had been very swift and, thankfully, free from suffering. There was a clerk bending over a carbonized ledger, pen still in hand, frozen in the immobility of death. Another was found bent over a wash basin, molded into a semistanding position. There were some in the streets who had managed to stagger a few steps. They lay on the ground with their bodies contorted and hands clutching at scalded mouths and throats. A family of nine was found still sitting around a charcoal breakfast table, their faces seared away. In all, the incinerated bodies of Saint Pierre's residents were found throughout the ruins of the city in every conceivable position.

When the charred bodies of the U.S. consul and his wife were found, both were sitting in chairs facing an open window that looked out on Mount Pelée. They must have seen the volcano explode and watched in awe as flaming death swept down and enveloped them less than a minute later. As the fiery cloud approached, the sight must have been so awesome that they never rose from their chairs. The consul seems to have had an intuition of death, because some months before the catastrophe occurred he told a visitor that he did "not expect to leave the island alive."

The May 8 eruption of Pelée was not a great one when compared to other volcanic outbreaks. The zone of destruction was less than ten square miles, but a set of circumstances made it rank among the most deadly in history.

The first factor was that it was so directional, and unfortunately the city of Saint Pierre lay precisely in its path. Secondly, from the moment the fiery cloud was ejected, it kept contact with the land until it reached the sea. Its extremely high density made it flow in concentration toward the city rather than losing some of its mass. Little of the gas was dissipated by rising in the air.

The final factor of destruction, and by far the most important, was that the city was not evacuated. The people were doubtless quite gullible, easily cowed, and willing to be herded by authorities and other self-serving influences. Newspaper editorials, extensive political double-talk, and mistaken scientific claims worked together to convince the townspeople that the volcano could not hurt them. Most of the "confused" people remained in Saint Pierre so they could vote in the upcoming elections—but none voted!

Amazingly, in the midst of this holocaust there were two survivors within the city. One was León Compère-Léandre, a shoemaker, who recounted the following story:

> On May 8th about 8 o'clock in the morning I was seated on the doorstep of my house. . . . All of a sudden I felt a terrible wind blowing and the sky became dark. I turned to go into my house, made with great difficulty the 3 or 4 steps that separated me from my room, and felt my arms and legs burning, also my body. . . . At this moment four others sought refuge in my room, crying and writhing in pain. At the end of ten seconds one of these, the young girl, fell dead; the others left. I then got up and went into the other room where I found the father dead. I went out and found in the court two corpses interlocked. Reentering the house I came upon the bodies of two

> men. . . . Crazed and almost overcome, I threw myself upon a bed, inert and awaited death. My senses returned to me in perhaps an hour, when I beheld the roof burning. With sufficient strength left, my legs bleeding and covered with burns, I ran to Fonds-Saint Denis, 6 kilometers from Saint Pierre.

Apparently Compère's home was located on the very fringe of the area affected by the fiery cloud. But this hardly explains why he should have lived when others were dying all around him. This is but one of many unsolved mysteries of this great catastrophe. The contrasts of the cloud's violence within limited areas were amazing. In the heart of Saint Pierre iron bars were twisted like pretzels, yet a short distance away fragile teacups were left unmoved and unbroken! Many groupings of bodies were found, some completely unclothed, while others in the same pile of bodies were scarcely disturbed and clothes were intact. One charred body was found next to a box of matches, none of which had been ignited. The fact that the fiery cloud can vary in violence within itself was a major factor in saving the life of the shoemaker. He lived in obscurity until his death in 1936.

The other survivor was Auguste Ciparis, a twenty-five-year-old stevedore who was imprisoned in the dungeon of the Saint Pierre jail. His crime was murder, for which he was scheduled to die on the morning of May 8. Instead he lived while all his keepers died. His cell had a single small opening on the door and no window, and this saved his life.

On the morning of his execution the prisoner was awaiting his free breakfast from the state. Instead everything got very dark, and a hot, stifling blast of air flowed through the grating on his cell door. He writhed in agony from the searing heat and fell to the floor. He later said he had absolutely no idea what had happened.

He lay on the floor for almost four days, suffering horribly from his burns and frequently calling for help. To add to his agony, there was no food and little water. He finally heard the sound of human voices above and cried out with what little strength he had left. Ciparis's cry was heard, and as soon as his rescuers established his whereabouts, they cleared away the rubble and freed the living dead man, the second and last survivor within Saint Pierre.

The doctors who later treated his burns doubted that he would live, but he did recover and lived until 1929. The new governor pardoned him for his crime of murder, considering that his four-day trauma while awaiting death or rescue had been punishment enough. Since Ciparis was now a free man, P. T. Barnum lost no time in signing him to his circus. He spent the next few years reenacting his ordeal from a spacious cross section of what represented his cell. The world could now see the survivor of Mount Pelée, billed by Barnum as "The Prisoner of Saint Pierre."

Mount Pelée's work was not complete; on May 20 a second eruption struck the city, but there was no more damage to be done. On August 30 it blew up again, this time killing nearly 2,000 more people in the neighboring towns. The most impressive thing to rise from Pelée was a huge mass of solidified lava. It grew straight up at the rate of about 30 feet per day, and by May 30, 1903, slightly more than a year after the catastrophe, it reached a height of 1,020 feet. Such a structure is known as a *spine*; this monolith was named "The Tower of Pelée." A photo taken at its maximum height shows the tower in the distance protruding from the peak of the volcano. Overlooking the ruins of the city, it seemed to serve as a memorial to the dead.

The tower was very brittle, and eruptions and internal unrest of the volcano caused it to fall apart rapidly. Great chunks of rock broke off the top and sides, and within a year the spine was reduced to a mass of broken debris. So ended the eruption cycle of 1902 to 1903.

Between the years 1929 and 1932 there were several more eruptions from Mount Pelée. On September 10, 1929, some minor explosions occurred, with ash once again falling on the streets of Saint Pierre. The people who had rebuilt and repopulated part of the city after the 1902 disaster needed no further excuse to depart. Saint Pierre is today a peaceful town, but it has never achieved the eminence of the pre-Pelée holocaust.

There has been little volcanic activity from Pelée since 1932, and a small museum has been established as a memorial of the tragedy of 1902. With "before" and "after" photographs of the city and artifacts preserved from the moment of the eruption, tourists are vividly reminded and natives will never forget. The story of the death of Saint Pierre is repeated regularly, often with the assurance that Mount Pelée is now an extinct volcano. However, because the zone is still tectonically very active, the volcano is far from extinct; instead it has entered a dormant stage. Its interior is slowly building up fiery molten energy while the outside of the volcano remains quiet and uneventful. In truth the volcano is merely asleep.

• •

CHAPTER SIX

The Wealth of the Gods

Since the beginning of civilization the minds of man have been filled with glittering dreams of gold. Of all the myths of gold the most compelling came out of the New World. To the Spanish explorers of the of the sixteenth century, gold was a prize for which they would sacrifice everything. For in the New World was located . . . the wealth of the gods.

One day in 1535 Sebastián de Belalcázar, veteran of the Inca conquest, founder of Quito, capital of Ecuador, met an Indian who fanned his imagination with a most unusual story. The native told the Spaniard of a king of a large tribe who lived near what is now Colombia. The nobles anointed him every day with a sticky gum and covered him entirely with gold dust. At the day's end he was carried to a nearby lake and rafted out to the middle along with an abundance of golden ornaments and emeralds. The Spaniard listened, fascinated, as the Indian described how the smoke from the many bonfires on shore drifted across the lake and the assembled multitudes shouted in awe and joy as their king slipped into the waters and washed off his golden "skin." Simultaneously the nobles and priests accompanying him on the raft threw the golden ornaments and precious gems into the lake as an offering to the sun.

Belalcázar had no reason to doubt the storyteller—had he not seen with his own eyes the fantastic gold hoards of the Incas? When the storyteller finished, the fired-up mind of the Spaniard coined a name that became engraved on the minds of fortune-seekers for centuries to come. He called the mythical king "El Dorado"—the Golden One.

When one considers the wealth of the Incan and Aztec empires prior to the arrival of the Spaniards, El Dorado could scarcely be entirely legendary. In their possession was far more wealth than the conquistadors could have imagined in their wildest dreams. Many scholars believe that the Incas alone mined more than two million ounces of gold per year and almost

twice that much silver. Although the totals cited above are possibly in excess, the amounts of precious metals mined had to be fantastic by modern standards. The Aztec output of gold and silver was only slightly behind that of the Incas. Doubtless, then, these two empires together possessed more precious metal than the combined treasuries of all the European countries. No wonder the invading Spaniards went berserk at the sight of so much gold!

These early Indians used gold mainly for ornaments and as a building material, in very much the same manner that contemporary Europeans used marble. There were life-size statues of people, animals, and naturally the gods, all molded from solid gold; such statuary was quite common. Although gold was sometimes used for bartering, it was not utilized as money.

The temples and, for that matter, all the important buildings were always roofed over with gold and, as often as not, decorated with golden friezes. Interior walls and pillars of the nobles' mansions were thickly plated with gold, while in their gardens solid gold fountains sprayed water into solid silver basins. Members of the nobility and their families would routinely dine from golden plates while sitting on golden stools. The nobles even played a game that strongly resembled modern bowling but with a slight modification. The balls and pins used in the game were made of solid gold!

Such vast wealth would be almost impossible to evaluate by modern standards, but certainly it would rival the United States' national debt.

Cuzco was the main city of the Incas; it was the microcosm of the empire. The first Spaniard to see Cuzco described it as "ablaze with gold." The Spanish king's inspectors, in written reports to their king, told of "buildings hundreds of feet long entirely plated with gold, in some instances almost a finger thick." From one building they took down 700 gold plates that together weighed 500 pesos of gold (1 peso = 4.18 grams). The Temple of the Sun, the residence of the sun-god Inti, was the greatest prize. The count of the gold plates that covered its walls has been lost, but the plates must have numbered several thousand, weighing from four to ten pounds each!

The most amazing feature of the Temple of the Sun was in the garden surrounding the temple. It contained a golden mimicry of plants. Maize, actual size, had been "planted" with its stalks cunningly wrought with gold in a soil of clods of earth, all of which were made of gold. All the stems and leaves of the garden were sculpted of gold. On the outskirts of the garden was thick golden grass on which twenty life-size llamas grazed. The llamas were also solid gold, as were the replicas of the shepherds who watched over them.

When invading Spaniards saw the cornfields of gold, they quickly became vegetarians and harvested the entire field; within two days not a single blade of golden grass remained. The defeated Incas could only stand by helplessly and watch. They probably expected Inti to exact his revenge, and when he did not, they must have concluded that even their great god was helpless before the might of the greedy Europeans.

The golden pieces of Aztec and Incan art were exquisite, but most of them quickly found their way into the Spanish melting pots. Only a few lovely ornaments survived. In 1931, however, an undisturbed Aztec tomb of a high official was discovered. The design and shape of the necklaces, earplugs, and rings, by their sheer intricacy and bulk alone, made one realize that the Spanish descriptions of the conquistadors' loot gravely

understated the rich artistic ability of the Aztec and Incan goldsmiths.

For centuries the Aztecs had been levying gold tribute from subordinate Mexican peoples, who were required to collect the gold by washing stream gravel in gourds and mining for it in the hills. Most of the yellow metal they produced was turned over to the Aztecs as taxes or tribute, for lowly people were not allowed to own the metal of the sun. As a result, gold accumulated throughout several hundred years of required mining and filled the temples and palaces of the Aztec capital Tenochtitlán (now Mexico City). It would be futile to attempt to describe such storehouses of treasure.

In 1520 Hernando Cortés marched on the Valley of Mexico with 600 men. His goal was the treasure of Montezuma II, the supreme Aztec ruler. Cortés had thus far been successful in his invasion of Mexico because of a superstitious quirk. The Aztec rulers, and especially Montezuma, mistook the bearded Castilian for a powerful god whose return to the Aztecs had long been prophesied. Fear of offending the god precipitated their destruction.

The Indians decided these white gods were too strong for their blood and wanted them to leave, peacefully. Montezuma sent rich gifts of gold, including a huge disk the size of a cart wheel that represented the Sun. It was of solid gold. The bearers of the gifts conveyed messages begging Cortés to turn back. As was to be expected, the sight of so much gold only served to spur the invaders on. Finally, with a great caravan laden with golden gifts, Montezuma himself ventured out to meet with Cortés. The great king was borne on a golden litter. This was the last great attempt to convince the Spaniards to take the gifts and leave in peace. Montezuma doubtless realized the futility of trying to distract the Spaniards from gold, because Cortés looked on such golden gifts as trinkets, mere samples of what lay ahead. And he was right.

The Castilian conqueror marched on Tenochtitlán. When the conquering Spaniards saw the wealth contained in the city, it was as if they were hungry sharks being fed raw meat—it drove them into a frenzy. It unleashed a fury that resulted in their committing some of the foulest deeds that so-called civilized man ever committed against his fellow men. The frenzy did not cease until the entire Aztec Empire lay in complete devastation, from which the survivors and their descendants have never really recovered.

In 1513 Vasco Núñez de Balboa and a small band of men crossed the disease-ridden Isthmus of Panama in search of Indian gold. Instead Balboa discovered the Pacific Ocean. He later tried to organize another expedition to explore South America's Pacific shore, but a rival for power managed to have Balboa arrested, tried for treason, and beheaded. One of Balboa's original band, an illiterate swashbuckler named Francisco Pizarro, went on to discover the golden kingdom of the Incas in a high valley of the Peruvian Andes.

With the coming of Pizarro, the Incan Empire was destined to suffer the same fate as the Aztec, as nothing could stop the frenzied greed of the conquistadors. The sight of the vast treasure of the Incas as described earlier was enough to seal their fate. The grandeur that was the Incan Empire was near its end; merely a shadow of it remains today.

The invading Pizarro rode inland in 1532 at the head of 368 men heavily armed with swords, crossbows and muskets. The supreme Incan king, Atahuallpa, who was worshiped as a divine offspring of the sun, came out to meet him with an escort of about 5,000 warriors. The king wore a string of large emeralds around his neck, complemented by a breastplate of gold that glittered in the sunlight. He was borne majestically forward, seated on a golden throne. None of his attendants were armed. Instead they wore precious gems and glittering golden ornaments. Doubtless Atahuallpa was making the same mistake Montezuma was destined to make; he showed the Spaniards too much gold.

Pizarro was even more deceitful than Cortés; he pretended he really wanted nothing more than their friendship. This was short-lived but effective, because it put the Indians completely off guard. Suddenly, without warning, Pizarro signaled his men to attack. The one-sided battle lasted no more than thirty minutes, and when it was over more than 2,000 Incan warriors lay dead and the remainder were taken prisoner, including the great king Atahuallpa. Historians recorded that not a single man among Pizarro's forces was killed.

The Incan king was held captive in a large room, where in desperation he offered the greatest ransom in history for his freedom. He agreed to fill the room, seventeen feet long by twelve feet wide, with gold as high as he could reach, which came to about eight feet. Accepting these terms, Pizarro drew a white line at this height. An adjacent room of much greater dimensions was to be filled with silver and precious stones. For nearly two months the king's subjects brought precious gold and silver artifacts into the rooms, and eventually the ransom was paid. The value of such a treasure, based on present market prices, would be hundreds of millions of dollars!

But treachery seemed to be a way of life for the Spaniard, because the king was not freed. To deprive the Incas of their leader, Pizarro had the lord Atahuallpa strangled in full view of his people. Poetic and impartial justice did triumph, however. A similar fate awaited most of the conquistadors, who were soon, and not surprisingly, fighting bitterly among themselves over gold and power. Pizarro was assassinated by his fellow Spaniards in the palace he had built for himself, in what is now Lima.

And thus ended the rich empires of the Americas. However, the vast hoards of precious metals obtained from the Aztecs and the Incas were but mere suggestions of the great metallic treasures that were subsequently mined and transported across the Atlantic to Spain. From Mexico southward, the Spanish conquerors established a vast network of mines worked, as a final outrage, by the Indians they had subjugated. The output was so great that convoys sometimes numbered over a hundred ships. Storms and privateers took their toll, but most galleons did make the crossing safely. Records show that during the 1550s alone the coffers of Spain were enriched by well over 100,000 pounds of gold mined from its possessions in the Americas. Who can dispute that the Wealth of the Gods had indeed passed into the hands of "civilized" man!

Word of the defeat of the god-king Atahuallpa quickly spread throughout the Incan Empire, and great hoards of gold were secreted away, remaining almost permanently elusive to the treasure hunters who sought them. The search still goes on, and quite recently one site has been located. It contains treasure from the two most sacred temples of the Incan religion, but recovery of that treasure does not seem imminent. Although how much treasure was actually housed there can never be known, all descriptions seem to be consistent. It is quite possible that much of what is described below really did exist. What is certain is that all the events included in this recounting are true.

In Lake Titicaca, Bolivia, are two islands that were once inhabited by hundreds of noblemen and priests. The Temple of the Sun, located on the large island, was reportedly crammed full of gold offerings. A gold chain 120 feet long hung around the altar. At least twenty strong men were required to lift it. On the altar was a huge sun-ray disk made of solid gold, which must have weighed almost a ton! There were numerous life-size replicas of large animals, such as llamas and jaguars, all made of pure gold.

To the east, about 500 yards distant, is the smaller island, which at that time contained the Temple of the Moon. It was said to be literally stuffed with silver offerings, statues, and an immense round silver shield, with a diameter of over ten feet, representing the moon.

Both temples housed numerous pyramidlike piles of precious gems, consisting mostly of emeralds and topaz. Some of these piles of wealth were over six feet high.

The network of spies established by the Spanish consistently reported

the contents of the temples to the conquistadors, who documented the information. When the invading soldiers advanced on the islands, they had great visions of limitless wealth. But the priests had their own spies, who reported the oncoming invaders. The priests, knowing what to expect, hastily dumped all of the temple treasures into the lake. The battle that followed witnessed the death of most of the islands' Incas. The conquerors burned the temples and destroyed the gardens, probably to relieve the frustration of losing such a gigantic treasure.

The gorge between the islands, where the priests dumped the treasure, is probably one of the deepest places in Lake Titicaca. So the treasure of the ages lies well over a thousand feet below the surface, buried under many feet of lake mud. To date all attempts at recovery have been futile.

The people who live near the islands today consist of a handful of impoverished Indian farmers. They scratch out a meager existence from the soil, without the slightest regard for the enormous wealth lying just offshore. But all is not completely lost; the ruins of the temples are still available to those who wish to visit them.

• •

"The Little Shop of Horrors"

Scattered throughout the world are over 500 species of plants that subsist partly on the breakdown of animal tissue. Such flora are designated as carnivorous or insectivorous plants since their most common prey is flying insects. They are usually found living in marshy places where little or no nitrogen can be obtained. The plants must obtain this necessary nutrient from the insects they eat. The major portion of their food, however, is provided through the process of photosynthesis, as is true for most of the earth's plant life.

Since plants are sedentary, they must develop a way to trap insects to obtain life-sustaining nitrogen. The various species have evolved ingenious methods for attracting and holding prey. Some carnivorous plants have

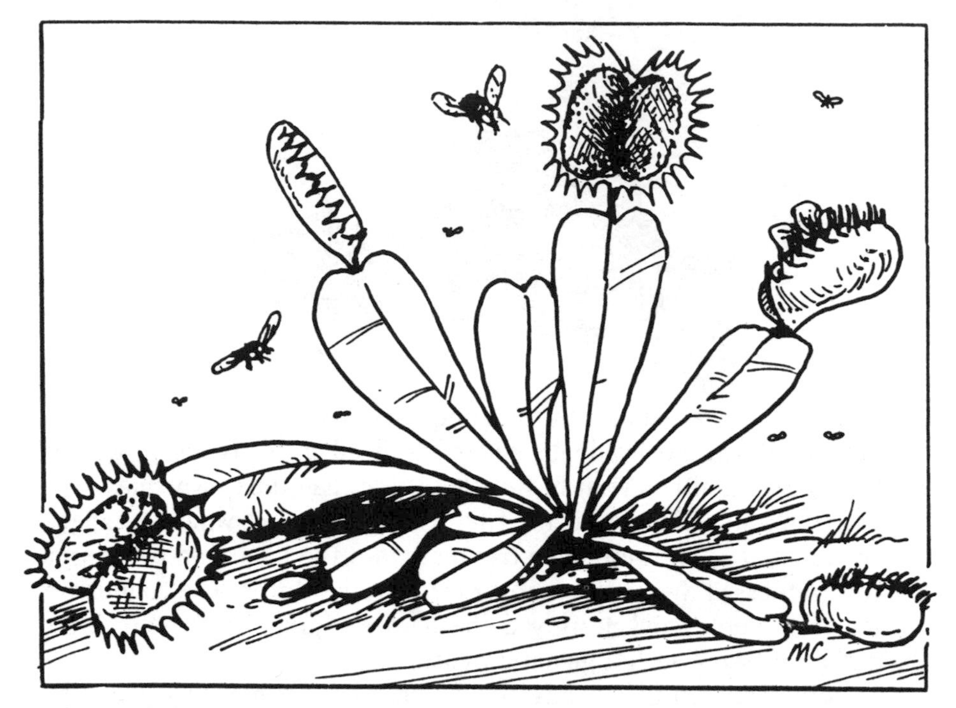

colored flowers that are scented in such a way that, at a distance, they appear and smell like decaying meat. This certainly attracts insects for dinner. Most of the plants contain glands that give off a digestive fluid, helping them assimilate their food. Nature has produced a number of remarkable species of carnivorous plants. One of the most extraordinary is also one of the best known, the Venus's-flytrap, *Dionaea*.

The Venus's-flytrap has evolved an almost perfect trapping mechanism. At the end of the plant's leaves are flaps that resemble an open bear trap. On each flap are three extremely sensitive trigger hairs. If an insect touches a hair only once, nothing happens, but if it touches two hairs or the same hair twice, the flaps spring closed and the sharp bordering spines interlock to imprison the victim. This closing motion is extremely rapid; scientists have timed some closing actions at less than half a second! The plant then secretes digestive fluid that dissolves the edible part of the trapped insect. In a few days the flaps will open again and reveal only the rejected carcass of the victim, which is soon blown away by wind.

This trap is so foolproof that, since two hairs must be touched or one hair touched twice, a windblown particle or a drop of rain landing on the open flap will not spring the trap!

Another carnivorous plant, also quite remarkable in its adaptation for catching prey, is the famous pitcher plant *Nepenthes*. It grows abundantly in the Malaysian forests and is related to a species common in bogs of the eastern United States. It is similar to the Venus's-flytrap in that the ends of its leaves have also evolved into an almost perfect trap. These leaves are shaped like large pitchers that are partially filled with water. The plant is brightly colored and gives off a peculiar odor. Both features serve to lure insects, which alight on the waxy lip of the pitcher. Because the lip is extremely slippery, the insect slides on it and falls into the water. Numerous downward-pointing hairs prevent the insect from climbing out. Eventually it is drowned and slowly digested with the help of a weak enzyme secreted by the leaf.

A most remarkably abundant carnivorous plant is the genus *Drosera*, more commonly known as the sundew. It is a relatively small plant with six- to eight-inch stems growing upward. Its leaves, about the diameter of a small coin, grow in clusters near the base of the stem. These leaves contain droplets of a sticky fluid produced by special glands. When sunlight shines on the fluid, it glistens like drops of dew, giving the plant its name.

It is this sparkling of light on the leaves that attracts the prey, usually a small fly. The leaves contain the necessary ensnaring equipment, in this case about 400 small, flexible hairlike bristles that are progressively larger toward the edge of the leaf.

When an insect alights on the leaf, the alluring fluid becomes the first snare, smearing and smothering the insect. The insect's struggles serve only to stimulate more secretion, and within minutes the inner hairs bend inward, gradually pinning down the captured prey. Very soon, after the outer hairs begin to curl inward, the insect becomes hopelessly engulfed in a web of sticky hairs. The edges of the leaf then curl inward and completely enfold the victim, which is by now past all signs of life. The glands then produce the juices that digest the insect.

Usually a week passes before the plant is satisfied and its meal is completely absorbed. It will then gradually unfold and reset the trap. The plant will once again patiently wait for its next victim to be attracted by the fluids on its leaves that glitter in the sunlight like drops of morning dew.

No discussion of carnivorous plants is complete without considering the subject of man-eating trees or plants. In fact, with the recent release of the stage hit *The Little Shop of Horrors*, interest in and speculation about man-eating plants have been renewed.

From time to time travelers to the island of Madagascar have returned claiming that a man-eating tree exists there. Strangely, though, whenever an investigation is undertaken, the tree is always unfortunately on another part of the island. In 1920 *American Weekly* published an apparent eyewitness account, with a memorable artist's rendering, of the sacrifice of a maiden to a carnivorous tree. Since the story, in a popular newspaper supplement, was widely accepted as authentic, the legend refuses to die. The truth about man-eating trees or plants is that they are not now, and probably never have been, in existence!

• •

Unusual Pets

In open lots throughout the United States children of all ages enjoy an occasional aerodynamic feat achieved by putting into flight and manipulating model airplanes. These are usually power-driven craft less than a foot long that are guided by movements of a string tied to the aircraft. A skilled operator can make it do limited aerial exercises, but usually the model aircraft flies in a continuous arc the length of the string held by the person guiding it. Enthusiasts have discovered many inventive ways to fly and control a plane, but one that has escaped their attention is quite common in equatorial Africa.

Here the children have discovered a way to amuse themselves with a similar pastime, except that instead of a model airplane, they use live beetles. A giant of beetles is the *Goliathus goliathus*, usually measuring over six inches long. It is quite capable of flight and makes a loud whirring sound as it flies. Native children search for specimens to use as pets and toys. They tie a long string around the beetle's neck and attach a stick to the other end. The beetle will fly in an arc the length of the string, making a loud whirring sound not unlike that of the toy airplane flown by American youth.

Beetle pets serve purely as a source of amusement, but in the Far East some practical uses have been devised for insect pets. Scientists generally agree that the praying mantis is among the most vicious of insect predators, but oddly enough, it can also be a friend to man. In Asia the mantis is highly prized and is often kept as a household pet. The owners often tie the mantis by a thread to their bedpost; in this capacity it serves as excellent protection from insect pests.

As always, some people are excessively indulgent with their pets. One elderly matron was so proud of her pet mantis that she had a silver collar designed for the insect and carried it on her shoulder, fastened by a tiny silver chain!

Others use insects as both pets and jewelry. Native women in Costa Rica secure luminous insects with tiny chains or cords and affix them to their hair or clothing. As the insect pets crawl about and flash their varicolored lights, they create a striking effect. Similarly, some women in Malaysia raise butterflies, using their pets as hair ornaments. The butterflies are extremely beautiful and display spectacular variegated colors. Tied to their owner's hair, they produce an unusual display as they attempt to fly about. The women are most attentive to their pets, each being identified by name.

An even more extraordinary custom has been practiced for centuries in West Africa. Here there are itinerant performers who capture and tame large scorpions, allowing them to live inside their voluminous clothing. The poisonous tail-spine is rarely removed, and yet these potentially dangerous creatures wander freely over their owners, even while they sleep on the ground. They never sting their hosts and rarely escape. They are truly pets and will respond to commands given them by a touch of their owners' finger. Really now, there must be an easier way to make a living!

• •

Avalanche—White Fury

An avalanche is a sudden cascading of snow and ice that occurs all too frequently on the steep slopes of the world's snow-laden mountains. It is one of nature's greatest destructive forces, easily ranking with tornadoes, hurricanes, earthquakes, and floods.

In mountainous areas where the slope is greater than thirty-five degrees, deep piles of snow often lie in a critical state of balance. A simple event such as the falling of a tree limb or a lump of snow, the weight of a skier, the noise of a pistol shot, or even the sound of a human voice can trigger a slide involving millions of tons of snow. In a matter of minutes entire villages can be smothered, causing inescapable death and destruction.

Initially the mass of snow moves very slowly, but it can quickly build up to speeds of over 200 miles per hour! In such a slide the frontal snow will lift off the ground, forming a whistling, swirling mass of powderlike mist. The powdery mist is often followed by huge solid masses of snow and ice.

The high velocity of a powder avalanche creates enormous air pressures in front of it, resulting in hurricane-force winds. Such a powerful wind once blew a tourist bus off a bridge in Austria, killing twenty-three skiers. Snow from the avalanche never even touched the vehicle!

The white fury of an avalanche is truly the skier's nightmare. Val d'Isère, France, was a popular ski resort until February 1970. Without warning, some 100,000 cubic yards of snow pulled loose from the adjacent mountains and smacked into the town at a speed of over 120 miles per hour. Within minutes forty-two people were buried under mountains of snow 300 feet deep. Survivors said they had the sensation that a tremendous explosion had occurred.

Dogs have been used extensively in the search for human victims of an avalanche. Stories of uncanny rescues attributed to dogs abound, many of which are true. One such event was recorded in Zurs, Austria. It was early in the morning; the local postman was busily delivering the mail. While driving the mail truck, he was so intent on negotiating the hazardous road that he failed to notice a blanket of white approaching. A sudden avalanche swept down, completely burying him and his vehicle. Although he was missed, nobody had any idea where to look for him. No one, that is, except his dog, which came to a certain spot and refused to move from it for three days. The persistent whining of the dog finally aroused the curiosity of the townspeople, enough that they decided to dig into the snow where the dog was keeping vigil. Here they uncovered the missing mail truck with the man still in the driver's seat, unconscious but alive. Almost miraculously, he survived having been entombed in the snow for over three days. Although the mailman recovered completely, the mail truck did not. Eventually it was sold for junk.

The worst avalanche disaster in the United States occurred at the Wellington Railroad Station, Washington, in the Cascade range. On March 1, 1910, a huge avalanche suddenly crashed down on the station. As its fury continued, it swept over a ledge and into a deep canyon three locomotives, several carriages, a large water tank, and the station house itself. Over one hundred people met their deaths.

• •

Medical Success Story

In 1900, when Dr. Walter Reed proved that yellow fever was transmitted by mosquitoes, he opened the door for a great medical success story.

In the 1800s the French attempt to build the Panama Canal was doomed to failure. Even with the most skilled engineers and plenty of labor at their disposal, the French failed to make much progress because of the terribly unhealthful conditions that existed in the region. The area was described as "one sweltering miasma of death and disease." In just eight years over 50,000 laborers died of malaria, yellow fever, and the plague. As the saying of the day went, one man died for every crosstie in the Panama railroad.

In May 1904 the United States took over the canal project from the French. The United States surgeon general appointed Dr. William Crawford Gorgas, an American army doctor, to tackle the problem of overwhelming disease in the Canal Zone. Dr. Gorgas had, in 1902, worked on eradicating yellow fever in Havana, Cuba. Within five months the disease, which had been widespread in the city since 1761, was under control.

Having already known from Dr. Reed's work that mosquitoes were the carriers of the dreaded diseases, Gorgas attacked them in every possible way, draining stagnant pools, spraying large areas of water with oil to kill the larvae, clearing undergrowth that sheltered mosquitoes, screening all living quarters, and dosing everybody with quinine.

His methods of combating the disease in infested areas were quite successful, and by the end of 1906 he had wiped out yellow fever and eliminated the plague in the Canal Zone. Malaria was more difficult, because the anopheles mosquito was harder to eradicate. However, by 1913, the area was relatively free of malaria. As a result of the tremendous victory man had won over an insect enemy, the canal was completed in 1914.

• •

When Worlds Collide

On August 10, 1972, our planet narrowly escaped collision with a meteorite. On that day a blazing white ball resembling a giant welder's

torch, trailing a long tail of smoke in its wake, soared over the western United States and Canada. It was a thirteen-foot object weighing at least 1,000 tons and flew at more than nine miles per second!

It was first spotted over Utah. A minute and a half later it was seen over Alberta, Canada. Had it collided with the earth, its punch would have been equal to that of the atom bomb exploded over Hiroshima during World War II. However, instead of hitting the earth, it virtually ricocheted off the atmosphere and flew harmlessly back into space. It seems rather startling that, after finding its way here across billions of miles of space, the meteorite missed hitting the earth by a mere thirty-five miles!

News of this near miss was considered so frightening that details of it were kept secret from the public for the next two years.

The collision of large meteorites with the earth could be devastating. Such an event occurred millions of years ago in the Shetland Islands, creating St. Magnus Bay. The bay is ten miles across and 540 feet deep at the center. Admiralty charts show the ridge around it forms an almost perfect circle. Scientists believe that the bay was excavated by the impact of a two-million-ton meteorite over 1,000 feet in diameter. It struck the Shetland Islands over a million years ago at a velocity of twenty-five miles per second—the ground must have indeed shaken.

More recently a large meteorite did collide with the earth, and it very nearly caused the destruction of Pittsburgh. On the evening of June 24, 1938, Pittsburgh, Pennsylvania, was rocked by a terrific explosion. Many people, seeing a brilliant flash across the sky and hearing the noise, naturally thought a powder magazine had exploded. The light and explosion were actually caused by the impact of an unusually large meteorite that struck beyond the western city limits. Had it fallen at a slightly different angle, it would have destroyed much of the city and killed at least half a million people!

However, this disaster would have been small indeed compared to a more recent near confrontation with an object from outer space. This close encounter occurred in March 1989, during the recent memory of practically every reader. Scientists detected the approach of a moderately large asteroid. It was well over half a mile in diameter and weighed many millions of tons. It did not come nearly as close as those listed above, but it did pass within 500,000 miles of the earth. As an encounter with a celestial body, this is considered very close, and scientists refer to it as a "cosmic near miss."

The destruction that would result were an object of this size to collide with the earth is unthinkable. If it had, for instance, hit the same spot as the Pittsburgh meteorite, the city of Pittsburgh would now be nothing but a gigantic hole in the ground!

• •

Quest for Fire

The discovery of fire by early man made living easier in the hostile world of the Stone Age. Fire warmed his cave in the winter, and because animals feared fire, he used it to stampede and bewilder them. It hardened his wooden spears, and cooking made his food surprisingly more palatable.

Lightning and volcanoes were probably the original source of fire. It is conceivable that early man was, at first, awed by a burning tree recently struck by lightning. As he salvaged a glowing ember, his fears were quickly replaced by curiosity. Man's intelligence suppressed his fright, and soon

fire was warming his cave. An old man or woman was probably assigned to keep the fire alive, as knowledge of how to start a fire artificially would not occur for aeons. If the fire watcher was negligent in his or her duty and the fire died, the wrath of the tribe would certainly have descended on so careless a keeper.

Until recently the earliest real evidence of man's use of fire was in the caves of Choukoutien, China. Here Peking man enjoyed the comforts of a campfire nearly half a million years ago. Scientists believe that, with his primitive intelligence, he probably did not make the fire, but instead brought embers into the cave and kept a fire burning constantly. One hearth in a Peking man cave revealed a pile of ashes twenty-two feet deep, testimony that *Homo erectus* had learned well the art of nurturing fires.

New discoveries in Africa seem to suggest that man's ancestors used fire nearly a million years earlier than previous evidence had indicated. In an archaeological dig in western Kenya masses of burned clay were found intermingled with a number of crude stone tools and animal bones. Chemical tests indicate that the clay was heated to around 750 degrees Fahrenheit, just about the right temperature for an open campfire. Radiometric dating shows this to be about 1.4 million years old.

A still older site, dating to about 1.5 million years ago, was recently discovered in a South African cave. Here two scientists unearthed a large number of charred animal bones. Many of them bore marks of cutting tools, indicating that the meat had been chopped off to be eaten. Cooking doubtless revolutionized the eating habits of *Homo erectus*, with meat becoming an attractive and nourishing addition to his diet.

Because theirs was a relatively hospitable world of mild climates year-round, the hominids (early humans) probably were not dependent on fire to keep their caves warm. Fire was a worthy ally, but its earliest uses would have been sporadic and dependent on when man was able to steal a flame from nature. Flame after flame must have gone out before he considered how he might preserve it or take it along with him on his wanderings.

Previous excavations in this South African cave complex show that a number of hominids were killed by large cats such as the saber-toothed tiger. Man may have discovered, as soon as he dared come close to nature's flames, that a burning stick could frighten away a predator. This discovery led to man's keeping a fire in his camp at night to discourage a cave bear or saber-toothed cat from entering. Then the possibility of hunting with fire may have occurred to him, and by igniting the forest floor he found he could flush prey into the open, where they would be at the mercy of spears, clubs, and stone axes. At about the same time he may have noticed that his wooden spear or bone or antler club could be hardened by exposure to fire.

Whatever uses the first men found for fire, scientists now know that at least 1.5 million years ago early hominids at Swartkran Caves, South Africa, found it an important key to survival. Two of the burned bones found at Swartkran are hominid finger bones. It is quite possible that some of the early humans may have been included in the cooking, but evidence of such ancient cannibalism is too slim to be conclusive.

With evidence that man used fire for at least 1.5 million years, it can safely be speculated that fire was probably first discovered about two million years ago. It seems amazing that so much time would pass before man was able to start his own fire by artificial means. This great achievement apparently did not occur until the time of Neanderthal man, whose tools included a fire starter, about one hundred thousand years ago.

The European Neanderthals were actually the first humans to have lived under subarctic and arctic conditions. In the summers they roamed a tundralike environment, following the various game herds. But during the harsh Ice Age winters such a practice became impossible. Armed with torches, they drove bears and lions from limestone caves and took the caves for their own shelter. While the winds howled and snowstorms raged outside the cave, they stayed behind windbreaks set up at the cave's entrance and huddled, comfortable and warm among animal furs, around central fires.

Many scientists believe that Neanderthal man did not become extinct, but instead evolved into modern man. Assuming this to be fact, perhaps it can also be assumed that, were it not for the Neanderthal's discovery of and ability to create artificial fire, these lines might never have been printed.

• •

Think Small

It is difficult to describe in meaningful terms just how infinitely small atoms and molecules really are. A single ounce of water contains approximately a trillion trillion molecules of H_2O (the chemical formula for water). If each water molecule in that ounce were enlarged to the size of a golf ball, all the ocean basins in the world could not contain them. Nor is the molecule the smallest building block in the division of matter, for it is itself made up of combinations of much smaller atoms.

There are many facts to dazzle the mind of those who are able to think small. The smaller a living creature is, the thicker will appear the liquid in which it must swim. To make this more comprehensible, the motion of a bacterium in ordinary water is comparable to that of a human swimming in liquid asphalt! However, since the bacterium does not know the difference, it is quite comfortable and adapts easily to its viscous environment.

It is difficult also for the nonscientist to maintain a frame of reference for the minute size of bacteria. They are the tiniest of free-living cells, so small that in one drop of water there may be as many as a million individual bacterial cells. Even with this vast population there is still enough space for them to move about quite freely in this single drop of water.

• •

The Year Without a Summer

The winter of 1815-16 was no different from any previous winter in southern Canada and the northeastern United States. Spring's arrival was normal; by April birds had returned from their wintering grounds and flowers had given color to the brown earth. But this was to become a most famous year, because it was the year without a summer.

April is often a cold month in the northeastern United States and southern Canada, but by May of 1816 the temperature still had not risen, and people became concerned. Morning after morning frost covered the ground as winter hung on.

On June 5 cold winds lashed the area, followed by a heavy snowstorm that covered the countryside with nine to twelve inches of snow. Freshly shorn sheep froze to death. The corn crop failed, and only the hardiest grains and vegetables survived. On June 6, at the inauguration of Governor William Plumber of New Hampshire, it was so cold, as one witness recalled, that "our teeth chattered in our heads, and our hands and feet were benumbed."

Weird weather continued into August with early morning temperatures always in the low thirties. On the few afternoons that were warm, people gamely tried to plant crops, only to have them destroyed by frost and snow. A killing frost occurred in mid-September; the new winter was slightly early, and it was to be unusually severe.

The spring and summer of 1817 returned to normal, and the weather has been predictable ever since. What caused the year of no summer? Many theories were proposed, but none even came close to the truth. A few scholars of the day suggested that an outbreak of sunspots had caused the chill.

The scholars of 1816 had no way of determining the actual cause of the extended winter, but the crackpots of the day presented explanations without hesitation. One outstanding theory of the soothsayers was that it had been caused by Benjamin Franklin. And of course several proponents of this hypothesis explained exactly why. The most commonly believed theory was that the hot interior of the earth releases heat into the atmosphere, but that because of Franklin's newly invented lightning rods, which were being installed all over the country, the earth's process of releasing heat into the atmosphere had been interrupted. This resulted in cooling of the air, and the summer season of 1816 was missed entirely.

Not all thinkers of the day accepted that theory, but a runner-up theory also blamed Ben Franklin. Its followers were firmly committed to the idea that, since lightning is heat, it must follow that the lightning rods had taken the heat from the air—hence no summer!

It seems ironic that as early as 1784 Benjamin Franklin had shrewdly speculated that dust from volcanic eruptions could affect climates by blocking out sunlight. He had made a connection between a constant "dry fog" in the atmosphere and the unusually cold winter of 1783-84. And he was right.

After many years of speculation and research, scientists now know what caused the year without a summer. The cause took place a year before and half a world away, in the Dutch East Indies. On the night of April 5, 1815, Mount Tambora, located on the island of Sumbawa, erupted with a force unmatched in recorded history. This gigantic eruption was even more powerful than the famous explosion of Krakatoa that was to occur sixty-eight years later.

Tambora ejected over twenty-five cubic miles of debris that blasted away nearly a mile off the top of the 13,000-foot volcano. It carpeted islands hundreds of miles away with layers of volcanic ash well over a foot thick.

The fine dust rose so high into the stratosphere that it encircled the world for years to come. The net effect was to screen out sunlight and thereby cause a drop in temperature, especially in New England and Canada.

The volcanic dust in the atmosphere affected other parts of the world as well as North America. Indeed it was an almost worldwide chill. In Western

Europe crop failures caused widespread famine, and many people starved to death. In Switzerland many people were reduced to eating Iceland moss and cats, and food riots broke out in France. Had this unseasonal drop in temperature continued for a few more years, continental ice sheets would have started to form and the earth could have slid into a new Ice Age.

A number of scientists predict that such wintry summers could occur again. Nature's volcanism and human industrial activity have caused a steady buildup of dust in the atmosphere over the last few decades. If this trend continues for about a century, it could produce an effect opposite to that of the greenhouse. World temperatures would be significantly lower and an age of ice would return.

• •

Fossil Caste System

Approximately one hundred million years ago an ant, about one-half inch in length, was crawling on a tree trunk when suddenly, without warning, it was overwhelmed and encased by a descending blob of resin. The resin eventually hardened into amber, thereby ensuring preservation of the ant.

In 1966 scientists dug the piece of amber out of ancient rock in New Jersey. The specimen, still containing the well-preserved ant, is now on

exhibit at Harvard University. A study of the fossil insect, scientifically known as *Sphecomyrma freyi*, showed the structure of the thorax and abdomen quite clearly. It was identified as a primitive working ant.

This was a rather important identification, because it indicates unmistakably that about one hundred million years ago, while dinosaurs were lumbering across the face of the earth, ants had already established their caste system of social organization!

• •

When Life Was Short

Life for man of the Old Stone Age was extremely harsh and precarious. Recently a scientific examination was made of nearly two hundred Old

Stone Age individuals whose age at the time of death could be determined. It was found that 55 percent of the Neanderthals and 34 percent of the Cro-Magnons died before reaching age twenty. Most of the remainder of those examined died between the ages of twenty and thirty, with fewer than 5 percent living beyond their fortieth birthday and three individuals reaching the ripe old age of fifty. Life for the female must have been exceptionally harsh, for practically all who survived beyond age forty were men.

The life span of *Australopithecus*, who lived about two million years before Neanderthal man, was even shorter. A physiological potential of up to sixty years has been projected for the *Australopithecine*, but few lived beyond their teens. Their mean life span of under twenty years testifies explicitly to the hazards of early hominid existence.

Scientists believe *Australopithecus* must have lived in rather cohesive family groups. Evidence is indirect, but considering that most of these early humans died during their teens, many of them certainly would have, at this age, left dependent children. Since this species of man survived for millions of years, the new orphans would have been reared by others, doubtless close relatives in the family group.

• •

Earth's Water Cover

At present well over 70 percent of the earth's surface is covered by water, and notably the large bodies of water are contained within specific basins. This is because of a peculiarity in the earth's evolution that resulted in there being continents and islands.

To grasp the magnitude of the earth's water cover, imagine all of the land leveled out evenly so that the earth's surface is a smooth sphere. With the waters distributed evenly over the entire world, the present supply of water would cover the entire globe to a depth of about 7,500 feet. This of course would make the earth a planet with a truly liquid surface that would appear from outer space as a blue planet with no surface markings and with widely scattered clouds.

In reality the average depth of oceanic waters at present exceeds five times the height of the land. The deepest sounding ever done, in 1962 in the Mindanao Trench, just east of the Philippine Islands, indicated that the water was well over seven miles deep.

To picture the relative depth of this area, imagine the tallest mountain in the world, Mount Everest (29,028 feet) dropped into the trench—it would still be covered by water to a depth of over one-and-a-quarter miles!

As stated earlier, over two-thirds of the planet's surface is covered by water. The volume is equal to about 370 quintillion gallons. Put still another way, there is enough water to provide each human on earth with about one hundred billion gallons—which is a lot of water. However, only a small and ever-decreasing fraction is available for human consumption. Use it wisely and sparingly!

• •

The Avocado-Loving Dog

The principal daily exercise for many dogs is running to their food bowl or begging for extras. More enterprising dogs scrounge meals from neighbors and local garbage cans.

Although basically carnivorous, numerous dogs can develop an appetite for almost anything, even candy and cookies. These junk foods, often given as hors d'oeuvres or rewards, supply far more calories than the average dog needs. The overweight dog is as much a problem to the veterinarian as its plump master is to a physician. The causes of canine obesity are similar to those of humans, with overeating at the top of the list.

Now a new problem has arisen in California: the avocado-loving dog. The avocado is a nutritious fruit containing many necessary food elements. It has a high caloric value and contains oils of unsaturated fatty acids.

The avocado has a definite place in the human menu but is an unnecessary addition to the canine diet. Some dogs, however, can't seem to leave them alone and have been observed to polish off all the fruit dropped in a grove. If that isn't enough, they will pick all they can reach from trees. Real dog gourmets even bury them to let them age. Some dogs carry an avocado around with them during the day in case they need a snack.

On such a diet weight gains have been phenomenal. One avocado-eating beagle, whose normal weight should have been about twenty-five pounds, weighed sixty-two pounds. It is an awesome sight to see a four-pound chihuahua devouring a half-pound avocado. Perhaps the owner should take a second look at his pet's dimensions before giving in to its wistful glance at guacamole.

• •

Galloping Glaciers

Because of the normally slow movement of glaciers, scientists often have to wait years to determine a change in glacial position. It is not uncommon for glaciers to show little change from one decade to the next. But occasionally, and for no apparent reason, some glaciers abruptly go wild in their movements. This is of much concern to geologists, who refer to them as . . . galloping glaciers.

Normal glacial movement, one or two inches per day, is almost imperceptible. But some glaciers move fifty to over one hundred feet per day. Such glacial surges, once considered a rarity, can occur anywhere that glaciers exist.

The precise cause or causes of these glacial advances are not fully understood. Documentation of normal glacial movements shows that all the ice does not move at an equal rate. Friction with the bedrock floor slows the ice movement at the bottom of the glacier, and drag along the valley wall slows the movement along the sides. One theory suggests that if there were sudden melting of a glacier along the underlying bedrock the ice would be released to move rapidly down the slope. Another hypothesis proposes that a block of unmoving ice at the terminus of the glacier may function as a dam. Eventually, pressure from the flow of ice behind it forces the stagnant ice to give way and propels the glacial ice forward.

The largest glacier in the USSR, the Fedchenko, normally moves about fifty yards per year, averaging five inches per day. Even that advance is relatively rapid for a glacier. Just a few years ago it unexpectedly surged and galloped at over fifty yards per day! This incredible pace continued for several weeks. Inhabitants of several villages had to flee from its path as the ice mass surged over their homes.

A rather famous surge was the Black Rapids glacier in central Alaska, which "galloped" in 1936. Near the highway was a hunting-fishing lodge that housed the caretaker and his family during the quiet season. Through the fall they constantly heard rumbling and felt the ground shake. They concluded, rather naturally, that it was an earthquake swarm. But they were wrong.

One December morning the caretaker's wife focused binoculars on a valley opening and shrieked an alarm. There she saw a gigantic wall of ice, a mile wide and over 300 feet high, bearing down on the lodge. As the ice moved forward, it cracked, throwing house-sized chunks of ice before it. They crashed to the ground in an icy spray, creating a terrible roar each time the ground was struck. The glacier was moving at the rate of over 200 feet per day, and the lodge appeared doomed. But less than one-half mile from the lodge the movement stopped, and it has never surged again.

At present the glacier is in a stage of retreating through melting, but enough remains as a grim reminder of the time a mountain of ice almost galloped over the lodge. This has remained a prominent tale of suspense in local history discussions and never seems to lack an interested audience. Many listen with confirmed incredulity—they would have to see it to believe.

• •

Man vs. Sea Mammals

Throughout many of the fishing grounds of the world a perplexing problem has developed between man and seagoing mammals. The struggle has become acute, principally because seals, dolphins, and some related species of whales hunt very successfully for the same fish sought by commercial fishermen.

The struggle has existed for decades, but with increased demand for seafood by humans the rivalry has become critical. In 1979 a group of Japanese fishermen herded over 1,000 dolphins onto the beach, where they clubbed them to death. They contended that a school of dolphins that large consumes over fifty tons of fish per day.

The competition for fish takes on many forms; along the Columbia River in Oregon fishermen closing their nets on salmon often find harbor seals darting in and grabbing their catch. Considering that a thirty-five-pound salmon is worth about $100 to the fisherman, this is quite a loss. The fishermen say the seals just relax in the sun, waiting for them to open the nets, and then in they go.

Punitive action such as that taken by the Japanese could result in mass extinctions of already threatened species. However, other forms of retaliation have begun, as numerous seal carcasses containing bullet holes have washed ashore in the Pacific Northwest.

In Antarctica several major fishing nations are planning to exploit the vast schools of krill, a two-inch shrimplike crustacean. This competition will inflict significant damage to some of the great whale populations, which now subsist entirely on krill. Each year the great whales consume almost fifty million tons of krill, even at their present reduced numbers. Removal of their food supply, which they share with seals, penguins, and squid, could spell the end for the largest creature that ever lived on the planet—the gigantic blue whale.

Scientists have inaugurated several programs of study designed to solve the problem, but as might be expected, they are proceeding slowly—much too slowly!

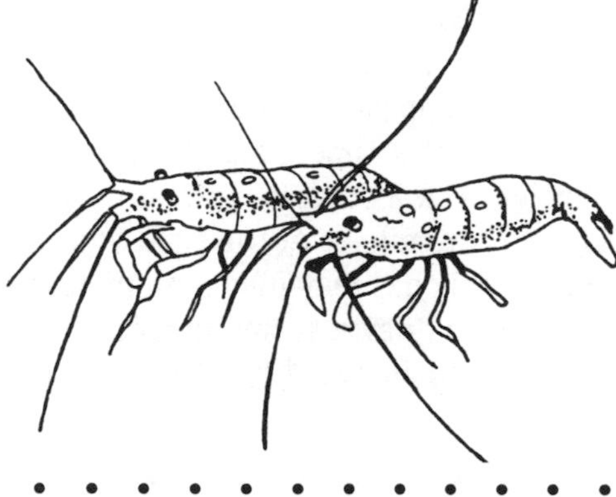

Tchaikovsky's Icebreaker Suite

Among people the whale might count as best friends, citizens of the Soviet Union hover near last place. That is, until the winter of 1985, when they redeemed themselves, at least in the minds of about 3,000 beluga whales.

In late December a pod of belugas (white whales) were chasing a shoal of cod near the Bering Sea. Unaware of the falling temperature, they lingered in the narrow Senyavina Strait too long. A brisk wind swept a wall of ice eleven miles wide and eighteen miles long into the passage, blocking their exit. There were only a few holes in the ice through which the whales could get to the surface to breathe.

When they were spotted by a native Chukohi hunter and his comrades, the hunters realized that this was too big a bonanza, and if they took advantage of this catch they would probably decimate the beluga population. Along with the villagers from the nearby settlement of Yandrakinot, the hunters decided to rescue rather than slaughter the whales.

Word of the trapped belugas reached Moscow, and soon helicopters brought marine mammalogists, reporters, and television crews to the scene. Local citizens dropped frozen fish to the whales, but their plight continued to get worse as the winter became harsher and the ice began to close in. The thousands of fifteen-foot adult white whales and their light gray young were packed so closely together that they had to take turns at the air holes. In typical whale parent behavior the adults would support the young as they gasped for air.

It was not until February that the icebreaker *Moskva* was able to come to their rescue. Twice the captain ordered the ship back when a huge wall of ice presented too much danger to the vessel. Finally, on February 22, the *Moskva* was able to plow through the twelve-foot-thick ice and open a channel to the unfrozen sea ten miles away.

The rescue effort didn't end there, however, because the whales didn't understand that they had been freed. Alarmed by the noise and vibration of the ship's engines and propellers, they refused to follow the *Moskva* into the opened channel, which, incidentally, was beginning to freeze over. The crew of the *Moskva* tried to accustom the whales to the ship by throttling back the engines, and in the meantime they worked at enlarging the pools for them to breathe. It was all to no avail, for they hid, then played around the ship, but still wouldn't follow.

It appeared that the whales would surely all die when one of the scientists had a brainstorm. He knew that whales were musical creatures, so he piped music over the loudspeakers. At first there were several false starts, but a few whales did begin to follow the ship as jazz and popular music were being played. When the music shifted to classical, more followed, and upon hearing the strains of Tchaikovsky's Nutcracker Suite

the whales all joined the audience and followed the ship through the channel into the sea and freedom.

Somewhere in the world of belugas, when music is discussed they swim transfixed by the memory of Tchaikovsky's music. They doubtless consider him their favorite composer, and no music can compare to his Icebreaker Suite!

The Unpoisoned Fruit

The most prominent citizen of Salem, New Jersey, challenged death before 2,000 spectators over 160 years ago. How? By simply eating a tomato!

It is difficult to find a modern home gardener who doesn't include tomatoes as part of his crop. Therefore, it's hard to believe that many of our ancestors wouldn't have allowed a tomato within sight of their kitchens.

The word *tomato* is a modern adaptation of the Aztec word *tomatl*. The invading Spaniards saw the *tomatl* growing in Montezuma's gardens in 1519. Although not much impressed, Cortés did bring some of the seeds back to Spain along with the more spectacular plunder. Tomato plants were soon growing in the gardens of Renaissance Spain. A visiting Frenchman described the tomato as *pomme d'amour*, or love apple, and it took on the reputation of being a mild aphrodisiac.

Along with its aura of being a love potion, the tomato was believed to pack a poisonous wallop. This was doubtless due to the first botanical description made in 1544 by Pietro Andrea Mattioli, an Italian herbalist. He linked the tomato to a number of disreputable relatives such as mandrake, henbane, and deadly nightshade. This association of the tomato certainly was a significant deterrent to eating one and was probably the reason for a 300-year intermission before the tomato became accepted as a regular ingredient in the human diet. Its reputation continued to be maligned, and finally, at the end of the sixteenth century, the tomato was nicknamed the "wolf peach," *wolf* because of its poisonous qualities and *peach* because it was deceptively luscious in appearance. This nickname, which in Latin is *Lycopersicon*, became the modern scientific name for the tomato.

Despite years of bad publicity, some Europeans did, without any ill effects, include the tomato in their diets and were doubtless willing to provide testimonials to its aphrodisiac qualities. In the American colonies, however, the people adopted the negative European attitude toward the tomato—and then some. Ministers and physicians generally denounced it for health as well as moral reasons. Pilgrims considered the tomato an abomination as shameful as theater attendance, dancing, or card playing. It is on record that one intolerably liberal pastor in New England was fired by his congregation because he grew tomatoes in his kitchen garden.

Thomas Jefferson, who was a talented, innovative gardener and an enthusiastic epicure, included tomatoes in his vegetable garden. He described them in his journal as a food plant and assured readers that they made an admirable preserve. Nothing much has been recorded about the reaction of the people to Jefferson's tomato patch, except John Adams's Puritanical reference to his spectacular cuisine as "sinful feasts."

On September 25, 1820, in front of the courthouse in Salem, New Jersey, the tomato was put on trial, and a giant step was taken toward its acceptance as an edible fruit. It was there and then that the courageous Colonel Robert Gibbon Johnson decided to prove to the world, or at least to the residents of Salem, that the tomato was not only safe to eat but also delicious and nutritional.

The man had probably discovered the truth about tomatoes while traveling extensively in Europe. He brought back plantings and encouraged local farmers to grow tomatoes as ornaments. Each year at the county fair he offered prizes for the largest and most attractive tomatoes. However, he never could convince anyone to take the first bite. The townspeople were understandably afraid of what the medical authorities had said the tomato would do. Reactions would be immediate, and death would be agonizing.

Determined to end this farce once and for all, Johnson publicly claimed he would eat a wolf peach. His neighbors were aghast. Even his physician,

Dr. James Van Meeter, declared, "the Colonel will froth at the mouth and double over with appendicitis."

Undaunted, Johnson appeared in public with a basket of tomatoes and said, "The time will come when this fruit will form the foundation of a great garden industry, recognized and eaten and enjoyed as an edible food.

"And now, to help dispel the scandalous tales, the wretched fables about this thing—to prove to you once and for all that it is not poisonous and will not strike you dead—I'm going to eat one immediately."

At the conclusion of that statement Johnson quickly took his first bite of a large, shiny red tomato. One woman screamed; another fainted. By the time he had finished his second tomato the crowd was cheering wildly.

The good doctor, Van Meeter, had been standing by from the very beginning. Expecting the worst, he was quick to realize his medical opinion had been dispelled. He snapped shut his medical bag and pulled an embarrassed but unnoticed tactical retreat down the street. All the while, Johnson was consuming the entire basket of tomatoes.

Shortly afterward, Americans began to eat tomatoes, and the colonel lived a long and healthy life to the age of seventy-nine.

• •

The Enemy of Mankind

Every continent has its share of dangerous animals, some a larger share than others. Of all the world continents, Europe contains the smallest number of species considered dangerous to man. In the past it was somewhat different, but at present the European bear and wolf populations have become so depleted by high concentrations of humankind that they claim very few, if any, victims. Europe can be considered even safer as one perceives that there is also a notable lack of highly venomous snakes throughout the continent.

The safest major country in the world, with no poisonous reptiles or spiders and no large predatory animals, is considered to be Ireland. In the entire country there is nothing more lethal than a bumblebee. But then, of course, there's man!

• •

Mosquito Blitzkrieg

Almost everyone alive today has at one time or another been bitten by a mosquito. In fact multiple bites are not uncommon, and they may be

inflicted anywhere on the earth except in the driest deserts, on the highest mountains, and in the most remote mid-ocean or polar regions.

Throughout history the mosquito has had a devastating effect on mankind. It bears responsibility for more discomfort and death than any other single form of animal life on earth. The malaria- and yellow-fever-carrying mosquitoes have probably killed more humans than all the wars in history combined. At least a million people still die from malaria annually, and untold millions are stricken.

In the tropics mosquitoes continue their war against mankind as vectors of dengue fever, filiarsis, typhus, African sleeping sickness, and elephantiasis. Among microscopic parasites that require a human host, there appears to be an "old girl network" that recognizes the need for a carrier to man. For this purpose the mosquito is perfect. It is the female that does the biting, as she must have blood for the protein needed to produce eggs. It is the bite that may release the disease-transmitting organism.

The victim usually doesn't feel that he or she is being bitten, but certainly hears the attacker, because her wings vibrate at a rate of 200 to 500 beats per second. The attack completed, she will fly off with about three times her normal weight in blood. The mosquito is now loaded with enough nutrition to produce up to 500 eggs, depending on the species. The mosquitoes must lay these eggs in water, and they will exploit every imaginable source of standing water.

During World War II soldiers stationed in the jungles of New Guinea were often horrified by what looked to them like a "solid wall of mosquitoes" on the trail ahead. All the men were issued mosquito repellent to smear on exposed parts of their body, but it was often ineffective because the repellent was quickly perspired off. These men must have felt that all the mosquitoes in the world inhabited New Guinea. While it is true that more than three-fourths of the species live in the tropics or subtropics, it would shock the men in the tropics if they could see the mosquito swarms that materialize in the Canadian Arctic.

There is a scarcity of species in the arctic and subarctic regions. In the short summer, however, thousands of square miles of flooded arctic tundra hatch mosquitoes in such enormous hordes that the sunlight is actually darkened. Life for both man and beast becomes unbearable at this time, and it is not uncommon to see workers in the outdoors draped in mosquito netting.

The mosquitoes seem to sense that summer is short and that they must move quickly to propagate their race. Their attacks on anything that moves are almost frenzied. The humming of millions of wings has reportedly driven persons insane. Many cases of the human mind succumbing to the drone of mosquito wings have been documented. Scientists in the Canadian Arctic have measured attack rates as high as 9,000 bites per minute. For an unprotected person this could result in loss of half of his blood supply in just two hours. This blood loss is more than enough to cause death by shock!

Scientists refer to these attacks as "mosquito blitzkrieg," and the term borrowed from World War II is most appropriate. To the infantry soldier in New Guinea it would indeed seem that the "solid wall of mosquitoes" had been transferred to the Arctic.

• •

It Happened in Manhattan

Rats directly or indirectly cause more damage in the United States than the combined depredations of all other animals. Millions of dollars' worth of stored food annually falls prey to rats.

The reproductive capacity of rats is absolutely remarkable. Scientists estimate that if all the offspring of a single pair of rats were to live and reproduce, over 350 million rats would be populating the area within three years! It is indeed fortunate for the human race that rats are relatively short-lived—those not killed will die of old age in approximately two years.

Because of their enormous and increasing population and their voracious, omnivorous habits, the destructive power of rats is escalating. They have become instrumental in causing fires by gnawing through insulation on electric wiring and by using oily rags and matches to build their nests. They have thrown entire cities into darkness by running across open switches and causing short circuits! A small consolation is that the rats causing such events are usually themselves quickly barbecued.

Rats can, and do, live anywhere: underground, in water, in trees, and particularly in the company of man. They are incredibly adaptable, and their lives have become closely linked with humans. They share our buildings, our possessions, and our food, attacking our crops before harvest, during storage, after preparation, and when it has been discarded. The living is easiest, they have discovered, near man.

Rats are quite at home in big cities; in fact the more densely populated the area, the closer it is to rat paradise. Here abundant food is practically theirs for the taking. They have adapted so well to man that human settlements are always thoroughly adequate for their needs. Because they have been partners with man for so long, the boundaries between host and "guest" have become blurred. They have reached a stage of open hostility toward man, as though they were the residents and we the intruders.

A near tragedy occurred recently in New York City. A man and his wife were startled out of sleep by the agonized cries of their infant son in the adjacent bedroom. Rushing into his room, they were horrified to find him bleeding badly from several head and shoulder wounds. On the baby's pillow a rat "nearly two feet long" was seated, staring defiantly at them. As the father tried to grapple with the rodent, it suddenly attacked first the father and then the mother. A vicious battle ensued, and eventually the man dealt the final blows with a club; but only after the rat had inflicted open wounds on both adults. The entire family required medical attention.

One might wonder, didn't they own a cat? If so, where was it? As a matter of fact, they had owned a large tomcat, but it had been killed by a rat just the week before!

• •

When the Coffers Overflowed

"And it came to pass in those days, that there went out a decree that all the world should be taxed."

Luke 2:1

The scourge of taxes has always been a fact of civilization, but the methods of ancient tax collecting were so diabolical that they would make the U.S. Internal Revenue Service seem like a charitable organization.

In ancient Egypt the wealthy were buried in elaborate tombs along with many of their earthly possessions on the supposition that they would take and enjoy all of their wealth in the afterlife. However, by 2500 B.C. the pharaohs had challenged the old belief that you can take it with you by taxing everything except the imagination—and no doubt would have included it if they could have found a way.

One ancient tomb, dated about 2400 B.C., has a series of reliefs depicting tax collectors at work. It shows the official sentencing of a lineup of tax delinquents and also illustrates some of the punishments. A standard penalty was flogging, and if the tax dodger did not change his ways and pay his taxes, as all good Egyptians should, he was thrown to the royal crocodiles!

Remarkable as it seems, the Egyptian investigators kept check on taxpayers very much as the IRS does today, by maintaining a tight record on individual property and income. Perhaps the IRS learned from the ancients. Among a number of tomb paintings that portrayed record taking, one showed a clerk off to the side, reed pen in hand, listing the size of the crop, recording meat cuts in slaughterhouses, and documenting production in workhouses. And the dry climate has helped preserve not only painted tomb records but also written records. Taxation and collection are preserved in facts and figures on numerous rolls of papyrus.

The greatest number of papyrus records that have been unearthed dates after 300 B.C., the time of the first of the fifteen Ptolemies and the beginning of Egyptian domination by foreign powers. The most powerful and enduring of the foreign conquerors were the Romans.

The Romans were even more efficient taxers than the pharaohs. One can say unequivocally that the inhabitants of the valley of the Nile owed their souls to the "company store." There was nothing more certain in life than the taxes they had to pay to their Roman conquerors. The Roman tax bureaus, at least as far as the Egyptians were concerned, were as complete and thorough as those of the modern IRS, despite the latter's sophisticated electronic instruments. They had a record of every single taxpayer, present and potential, and managed to tax everything imaginable, with collection methods characterized by ruthless efficiency.

House-by-house registration kept a running record on all occupants.

These records included such data as age, profession, tax status, and property owned. Official tax clerks saw to it that all the rolls were kept up to date. Land registration encompassed all farms and orchards, including the likely value of the yield. A tax was levied on each individual cow, sheep, and pig. Craftsmen and tradesmen registered with the tax collector all their craft and trade goods. Even prostitutes had to register their nightly take so that it could be taxed.

Tax offices were absolutely jammed with record ledgers of past, current, and future taxes. A portion of one record survived. It documents taxes levied against a small village for a three-year period. Scholars note that this single remnant of a tax record is as long as Homer's *Iliad*.

Taxpayers were issued receipts for the taxes they had to pay. The one major difference from the modern IRS is that they had to supply their own writing material. To save the cost of paper most taxpayers instead scribbled their receipts on chunks of broken pottery.

Failure to give Caesar his due resulted in very stiff penalties. However, since flogging was no longer practiced, local toughs were hired to collect delinquent taxes. They would break into the house of a tax dodger and clean out everything in sight. If the landlord or his family protested, they were either beaten up or, as in many cases, sold into slavery. These tax raids occurred frequently, because the salary of the collectors consisted of a percentage of the take.

The methods of ancient tax collectors discussed above are in no way meant as suggestions or encouragement to the "untiring resources" of the United States Internal Revenue Service.

• •

CHAPTER SEVEN

Iguanodon

In 1841 Sir Richard Owen presented a scientific dissertation at the meeting of the British Association for the Advancement of Science, which convened in Plymouth. His paper was published the following year in the proceedings of the association. Owen was describing a new suborder of saurian reptiles ". . . to which I would propose the name of Dinosauria."

The first dinosaur to be named and described scientifically was identified by Dr. William Buckland in 1824. His description was based mainly on a jaw of a carnivorous dinosaur found in the early nineteenth century at Stonesfield in Oxfordshire. He named it *Megalosaurus*.

The second dinosaur, *Iguanodon*, was named and described in 1825. It is one of the best known of the dinosaurs, because no fewer than thirty-one adult skeletons had been found in close proximity, all of which had died simultaneously as they stampeded off a high cliff.

The initial discovery of *Iguanodon* was made on a spring morning in 1822, when Mary Ann Mantell was strolling along a quaint country road near the small city of Lewes, south of London. She had accompanied her husband, Dr. Gideon Mantell, to the house of a patient, and while he was inside she wandered down the road enjoying the beautiful day. As she walked, she passed a pile of rocks that had been dumped near the road's edge to be used for repair. Among the rocks were some objects glittering in the morning sunlight. Curious about them, she picked them up and found that they were large fossilized teeth.

Mary Ann was absolutely right in believing that her husband would be interested in her find. Dr. Mantell was no ordinary medical practitioner; he had an intense interest in fossils and studied them assiduously. As she suspected, when Gideon emerged from the patient's house and was shown the teeth he became very elated and immediately recognized them as the teeth of some unknown plant-eating animal. Although the word *dinosaur*

would not be in use for another twenty years, the animal came into scientific focus at once.

Dr. Mantell researched the quarry from which the fossils had been taken and managed to obtain more teeth and bits of bone. The quarry was located near Cuckfield, in the Tilgate Forest. His immediate assumption was that they represented a long-extinct colossal reptile, but he had nothing more in the way of evidence. Unsatisfied with the response of his scientist friends, he submitted the fossils to the great Baron Cuvier, who was considered the founder of comparative anatomy. The great one identified them as the teeth and bones of a rhinoceros, reaffirming that greatness is not synonymous with accuracy. When Mantell later sent him some ankle bones from the same quarry, Cuvier identified them as belonging to a species of hippopotamus.

The casual identifications did not reassure Mantell, so he took the bones to London to work at the Hunterian Museum at the Royal College of Surgeons, where was housed an excellent collection of bones. Here he met a naturalist friend, Samuel Stuchbury, who had just returned from South America. When Stuchbury saw the teeth, he showed the doctor an iguana tooth, and they noted the resemblances. This was sufficient evidence for Mantell, so he assigned the second scientific name to a dinosaur, *Iguanodon*, meaning literally "iguana tooth."

During the next fifteen years other bones of extinct reptiles were found, and in 1834 the partial skeleton of an *Iguanodon* was found at Marlstone in Kent. Dr. Mantell purchased the skeleton from its owner for the price of £25. With the fossil remains on hand, Mantell did try to reconstruct the animal. But in attempting to match it with a modern lizard, he ended up with nothing more than an interesting arrangement of bones.

The first official reconstruction of *Iguanodon* was displayed publicly at London's famed Crystal Palace exhibition in 1851. Sir Richard Owen and Waterhouse Hawkins were responsible for the life-size *Iguanodon*. Their assemblage of bones resembled an oversized reptilian rhinoceros with a horn on its nose and a skin covering of scales resembling tiles on a bathroom floor.

Iguanodon lived during the time when much of Europe and North America enjoyed a tropical to subtropical climate. The animals lived in herds and were apparently abundant. A full-grown specimen was elephant-sized, weighing about four to five tons. It was able to rear up to a height of about fourteen feet, and was probably bipedal. It browsed in trees much of the time but could crouch on all fours to graze amid ground vegetation. It ran on two legs with head and body leaning well forward, balanced by the outstretched tail. Its forelimbs were much smaller than its hind legs, and on each thumb was a large, bony spike. This spike, the same one that Sir Richard Owen had placed on the nose of his reconstructed model, was probably used as a defensive weapon.

One American species appears to have been much larger than the elephant-sized European *Iguanodon*. In 1937 a series of *Iguanodon* tracks was hewn out of the roof of a Colorado coal mine. The prints, which were about three feet long, were impressed at intervals of fifteen feet, certainly indicating a creature of great stride. This animal was estimated to have been at least thirty-five feet tall, but the animal itself continues to be an enigma, since no skeletal remains were ever found.

To establish the setting for the *Iguanodon* story, one must go back nearly 300 million years in the country of Belgium, during a time European geologists call the Carboniferous Period. The land in Belgium was heavily covered with forested muddy swamps. As the trees died, they fell into the swamps and were buried naturally. With the passage of time the accumulation of buried vegetable matter turned into peat. The weight of millions of years of subsequent deposits, coupled with steadily increasing pressures, gradually transformed the peat into coal.

Now it is necessary to race ahead in time to about one hundred million years ago, when dinosaurs dominated the earth. This is the period known as the Cretaceous, and it was the time of the *Iguanodon*.

Erosion had cut a ravine at least 200 feet deep into the Belgian soil and rock during the intervening 200 million years. Peering across the ravine from one rim to the other side, one could readily discern the old Carboniferous coal seam on the cliff face. Erosional processes had cut right through the ancient coal deposit. It was into this deep ravine that no fewer than thirty-one adult *Iguanodons* had fallen in quick succession, lodging at various levels on its lower slopes. With the passage of time the ravine underwent numerous floods and was gradually filled with Cretaceous muds that buried the dinosaur skeletons and in time completely filled the ravine.

In 1878, in the Belgian coal mine at Bernissart, a most extraordinary discovery was made. Some coal miners who were developing a new gallery encountered numerous fossil bones at a depth of over 1,000 feet. They had picked their way almost completely through one skeleton before becoming aware of its presence. A noted paleontologist was called in to survey the situation. He recognized the bones as belonging to *Iguanodon* and observed that they were present in great numbers. Excavation then proceeded in earnest. A scientist-engineer, M. De Pauw, spent the next three years directing the work of digging out the skeletons.

At first they tried to excavate under the fossil graveyard by digging another tunnel over a hundred feet below the original find, but again bones were encountered. This dinosaur cemetery was evidently of gigantic proportions, with its vertical extension through more than a hundred feet of rock. It became obvious that the bones were not contained within the stratified beds of the mine, but instead were embedded in unstratified clays of a later geologic period. These filled a cut through the layered coals and shale. In short, the *Iguanodon* bones were distributed in what had once been a deep ravine, cut into the older Carboniferous rocks. The bones were embedded in the unstratified clays that had in time filled the ravine.

The meticulous mapping done in the mine was coordinated closely with the excavations of the fossils and revealed the detail of the ancient landscape described above. The outline of the banks of the Cretaceous ravine was delineated by the inwash of sediments that filled it. These sediments were quite distinct from the regular stratified layers of coal and shale being mined so far beneath the surface of the earth.

The reconstruction and mounting of the recovered skeletons was done mainly under the direction of Louis Dollo, who devoted his entire career to the preparation and description of *Iguanodon*. He mounted the first skeleton in 1883, and by the turn of the century four more had been erected, all in life positions. At present the Royal Museum in Brussels houses thirty-one specimens, eleven standing skeletons with twenty additional complete and partial skeletons recumbent at the base of the exhibit.

This is one of the most sensational dinosaur collections anywhere in the world. It has been invaluable for research because the incredible menagerie mustered from one particular site allows great scope for comparative study of variations within a single species of dinosaur. Size differences within a variety of specimens, if found separately in different locales, can and often do lead to their being classified as separate species. But here, where they can be compared with each other, it can be seen that the variations are really sexual and age distinctions.

Excavations were stopped in 1881, although more bones were observed. During World War I, however, when Bernissart was occupied by the Germans, a scientist was assigned to investigate the reopening of the mine and the fossil deposits. The work did not advance, because the area was reoccupied by Allied Forces, who had other things to occupy their attention. The site was completely abandoned in 1921, after which it rapidly flooded. It is impossible to say just how many more skeletons of *Iguanodon* lie buried in this dinosaur cemetery. The specimens that were recovered doubtless represent only a small percentage of the herd of *Iguanodons* that stampeded off the cliff's edge so many aeons ago.

It is not difficult to re-create the scene as it happened one fine Cretaceous day about one hundred million years ago. Imagine, to begin with, a herd of *Iguanodons* feeding on a large patch of vegetation overlooking a deep ravine. Although absorbed in their meal, they kept a wary eye on the nearby jungle, knowing it could easily shelter a hunter. Their worst fears were realized when suddenly two monstrous megalosaurs charged out of their forested cover and sank their teeth and claws into a large *Iguanodon* that had wandered too close to retreat. The helpless animal howled in pain and terror, causing an immediate reaction from its fellow *Iguanodons*. Their behavior was essentially no different from that of present-day prey animals reacting to the charge of a pride of lions.

Almost instantly the entire herd of *Iguanodons* turned and stampeded away from the scene of the attack. At full speed all ran in the direction of the deep ravine, and in their undefinable terror the entire herd stampeded off the cliff that lay to the south. Briefly the air was filled with cascading dinosaurs, turning and twisting, striking the slopes of the ravine about one hundred feet from the top. Here their broken bodies continued the downward trip, rolling and tumbling until they finally came to rest either somewhere along the slope or at the bottom of the 300-foot ravine.

Not all were killed by the fall, and the luckier few hit the gentler incline of the ravine's slope. These survivors were doubtless badly bruised as they fell to the bottom. They were, at least, in no danger from the megalosaurs as they struggled painfully to their feet and limped away, living to run another day.

Most of the herd fell where the incline of the cliff was quite steep, and they died on impact. A few battered specimens may have tried to get to their feet and move away, but broken bones cannot support a four- to five-ton body. When the dust cleared, the bottom of the ravine and the north slope were virtually covered with the broken bodies of a herd of *Iguanodons*, some still waving their heads feebly as they breathed their last.

On the flatland above the ravine the two megalosaurs, alone with their victim, began to feed.

• •

The Born Losers

The earliest recorded solar eclipse occurred on October 22, 2137 B.C., as documented in the Chinese classic *Shu Ching*. The ancient Chinese believed that the sun was constantly under attack by dragons that took large bites from it. The confirmation of this, to which they might point to verify their claim, would be a partial eclipse with a missing bite. Most emperors appointed at least one court astronomer whose sole duty was to predict an eclipse. Because the emperor was forewarned, the soldiers could rescue the sun by shooting arrows into the sky. The dragon was always killed or driven off. Nobody could dispute this, because within a short time the sun began to shine again.

Predicting eclipses was not as easy as it might seem, because astronomy was quite a dangerous profession in those days. The emperor invariably decreed that if the court astronomers were negligent (or incorrect) in their predictions, they would suffer instant death!

A little over 4,000 years ago, during the reign of Chung K'ang of the Hsai Dynasty, the emperor appointed two men, whose names were not recorded, to the post of court astronomer. The two men were quite successful for a while; in fact, whenever they felt it was necessary to prove their prediction prowess, they would announce an impending eclipse. The emperor's warriors, thereby alerted, would fire arrows into the sky and beat gongs and drums and thus frighten away or kill the dragon. The sun was always rescued, and the eclipse was avoided. On one fateful occasion, however, the two astronomers missed a guess and failed to predict an eclipse. Although the warriors were able to "rescue" the sun and it returned shortly, the astronomers were executed for their oversight.

All was not lost, as they left behind the first actual record of an eclipse.

No one thought to ask them to turn over their secret formulas showing how they were able to predict eclipses. Possibly their failure rendered their methods too unreliable to preserve.

Through the years a heavenly eclipse, particularly of the sun, has held intense fascination for professional and amateur astronomers. Many would go to great lengths to witness and study such events. Some of the results, had they not been so frustrating to the scientists involved, would have been more appropriate in the comedy category.

Many astronomers were absolutely fanatical about the subject. Foremost was a Scottish scientist who lived during the days of sailing ships. In trying to observe six separate eclipses, he actually chalked up a sailing record of 75,000 miles to gain the best vantage points for observation. Despite all his efforts, he observed only one, the other five being obscured by cloud coverings. As a consolation it could be said that he had the opportunity to see the world, if he bothered to look.

His experiences were perhaps no less frustrating than those of Pierre Janssen, a French astronomer who was absolutely determined to photograph the solar eclipse of 1871.

His was a difficult situation, because he was in Paris while it was under siege during the Franco-Prussian War. He secured a hot air balloon and, risking heavy German rifle fire, rose above the city. The Germans were so startled by the sight of the balloon that not a shot was fired, and Pierre escaped unharmed. Without losing time, and after undergoing considerable hardship, he traveled by whatever means possible to his observation site. He finally reached the path of the eclipse on the East African coast well in time to observe it, but the event was obscured by rain!

• •

Romance by Odor

Communication among many animal species is limited, and so, in order to help, nature has devised a set of unmistakable sex signals. One of the simplest and most important is the odor that the female of many species emits when she is ready to mate. Response by the male is immediate.

It seems ironic that some of these natural fragrances, such as musk and civet, are the bases of many of the perfumes in great demand and use by modern women. They are used in a manner similar to that which nature originally intended, and as a consequence men are attracted to the pleasant smell a perfumed woman emits. Therefore the well-groomed woman who is wearing a perfume based on the essence of musk should not be frightened if she finds herself being followed by an alligator. It just happens that musk is the sexual aroma that turns him on. His intentions are purely romantic.

• •

So That the Race May Live

Every species of animal life is remarkable in its own way, but those that astound us most are the ones whose behavior is not directed at satisfying their own immediate needs. For some animals instinctive behavior often involves extreme hardship and self-denial, merely to provide the best possible odds for the next generation. Such creatures deserve our admiration; the salmon is one of them.

Salmon will swim great distances, often against violent currents, to spawn in the precise stream in which they were born. This group of fish matures in the ocean, where some individuals may grow to lengths of four feet. At the appropriate season they seek out fresh water and begin a most frenzied swim upstream. They buck the swift currents of rapids and leap waterfalls as high as ten to fifteen feet. Considering that salmon may weigh more than seventy pounds, this is quite a feat. Nor does their skill and prowess stop there; the fish must constantly defy the waiting carnivores, usually bears, who rank salmon among their gustatory favorites. Throughout their frantic swim upstream they never pause, not even for food. The digestive system of the Pacific salmon degenerates, and they never eat again. During this incredible journey they change appearance as their silvery scales turn into a characteristic red.

For the Pacific salmon the journey ends when they find the stream in which they were born. But the most amazing salmon are those that swim up the Yukon River. The fish usually swim the entire length of the river, over 1,000 miles, seeking the waters of their birth. Experimentation has shown that the salmon are able to locate their birth stream through an extraordinary sense of smell! When the Yukon salmon reach their destination, the eggs are laid and fertilized. Their task in life now completed, the noble fish simply retire from life, never to see their young or to return to the sea. The newborn salmon swim downstream to the open sea, where they will mature; thus the cycle of life is repeated and the race lives on.

The Atlantic salmon undergoes the same frenzied swim upstream to its place of birth. But unlike its Pacific counterpart, the new parents do not die; instead they return to the sea and resume whatever ventures they normally pursue. The young eventually follow their parents to the sea. There they will mature and repeat an instinctive cycle that seems to defy the very basic laws of survival.

A number of animal groups besides the Pacific salmon instinctively make the supreme sacrifice to preserve the race. Among them are several species of the female octopus. Nature has programmed her to self-destruct after she lays her eggs; it is then that the optic gland secretes a substance that induces appetite loss. Therefore, instead of hunting for food, she spends her last month of life constantly protecting her nest. In this manner her brood is assured of a good chance of survival—while the mother calmly starves to death.

• •

The Night the Sky Fell

Watching the sky on a clear, flawless night, one should see a "shooting star" about every ten to fifteen minutes. The shooting star is really a fragment of comet dust that, upon entering our atmosphere, encounters friction, causing it to burn up. For the most part these individual particles of meteoric dust were once part of a comet.

The nucleus of a comet is not really complex. It consists mainly of frozen gases with many bits of solid matter, all of which are weakly bound together. The binding is so relatively weak that a comet is easily pulled apart when approaching a large celestial body. Therefore, since most known comets have an orbit that stretches from one end of the solar system to a point close to the sun, their life span is, astronomically speaking, short.

When a comet approaches the sun, many pieces of the comet are, in essence, torn away by the gravitational influence of the greater celestial body. These pieces continue to move along the path of the mother comet's orbit. An entire comet may in fact eventually be destroyed in this manner, but the remnant comet dust particles, which could number in the billions, continue to move through space in the original orbit. When the earth passes through one of these dusty trails, the comet debris enters our atmosphere and a meteor shower results.

The Biela Comet, discovered in 1827, had a very predictable periodic orbit that crossed the path of the earth quite punctually every six-and-three-quarters years. When the comet came into view in 1846, scientists were astonished to find that it had split in two. The two halves traveled like celestial sisters, each with head, nucleus, and tail. They reappeared for the last time in 1852, and since they were not seen again the Biela Comet was presumed lost.

The comet had apparently come too close to the sun, initially causing it to split in two. Thereafter the gravitational pull of the sun was so great that both halves fragmented. The countless particles of what was once the Biela Comet continued along the same orbit. This was a trail of celestial dust that crossed the orbit of the earth on November 27, 1872, at about 7:00 P.M.

As the cloud of particles enveloped the earth, friction with the atmosphere produced the most spectacular meteor shower on record. The sky was crisscrossed with bright crackling streaks that "fell like snow." It was estimated roughly that over 160,000 shooting stars flashed across the sky that night. To the superstitious it must have seemed indeed that the sky was falling.

• •

By the Skin of His Teeth

Much of the fossil mammal collection in the American Museum of Natural History in New York City was gathered by Edward Cope in the American West during the time of the Sioux Indian wars.

Being a Quaker, Cope refused to carry a gun, despite the obvious dangers he would encounter. Without realizing it, he apparently was equipped with a built-in weapon. On one occasion his fossil hunting was interrupted when he suddenly found himself surrounded by hostiles. His quick thinking and action undoubtedly saved his life and that of his associates. To confound and amaze the Indians he simply removed and replaced his false teeth over and over again. The Indians were so bewildered that they left without harming anyone.

Such stories do not end with Cope, for expeditions into untamed land will produce legends. A rather tenacious tale about another scientist persists in the annals of the American Museum of Natural History.

During the 1920s this museum conducted a number of field expeditions into the Gobi Desert under the leadership of Roy Chapman Andrews. In 1925 Nels C. Nelson, at that time curator of prehistoric archaeology, was a member of the Central Asiatic Expeditions, as they are known. His outstanding idiosyncrasy was his glass eye.

During the 1920s Mongolia had undergone a revolution and was a wild and lawless area ruled by warrior princes and crisscrossed by nomadic bands. At one point the expedition ran up against a hostile and well-armed group of Mongols on horseback, all with the usual cartridge belts slung across their chests.

The warriors dismounted and immediately began to harass the nervous expedition leaders for favors. However, Nelson kept his cool and was struck by what proved to be a great idea. He pretended to be a powerful magician and removed his glass eye. Holding it high for all to see, he quickly slipped it back in its socket. The Mongols fled.

• •

Bird Mercy

Almost daily, aircraft somewhere cause fatalities among birds in flight. The worst time, of course, is during the migration season, when the number of birds killed by aircraft undoubtedly reaches the thousands.

There was, however, one newsworthy incident in which aircraft were responsible for saving the lives of several thousand birds. It was in the fall of 1931 when an early blizzard swept over Austria and prevented thousands of swallows from migrating south. As a result they began to die from the cold and the lack of food.

Fortunately Vienna was home to a number of bird lovers at that time. They managed to charter two aircraft to fly some of the swallows to their winter destination, their migration flight having been canceled on account of the weather. Over 27,000 swallows were collected in and around Vienna and flown across the Alps to Venice. Here they were liberated under sunny skies in a temperature forty-two degrees warmer than that of Vienna when they left. They spent the winter in Italy and prospered.

For humans this was an unusual deed of mercy, for it is too common for us to remain unaware and uninvolved. Birds therefore must occasionally be prepared to apply their own methods of compassionate behavior. Perhaps the crow, whom many would describe as merciless, has some surprises in store for scientific observers. Recent observations have shown that when a flock of crows perceives that one of their number appears dejected, they will sometimes gather around the depressed bird in great numbers. In an attempt to revive its spirits they will render a chorus of continuous, very loud and lively cawing. This morale-building set of tactics often works, and the dejected crow appears to respond in a positive manner.

If, during the loud cawing, the dejected bird remains despondent and seems unaffected by the treatment, the flock of healthy crows will attempt to relieve it of its misery. Their method is to attack the unresponsive, miserable bird and peck it to death. Scientists consider this to be a form of mercy killing among crows, which sense the depressed one is dying.

• •

Solomon's Treasure Island

The country of Sri Lanka, formerly Ceylon, is a tropical island about the size of Ireland. The unique feature of the country is that it sits on an immense treasure. Gem-quality minerals can be found almost anywhere. In fact Sri Lanka is in the top rank of countries that produce gemstones, particularly rubies and sapphires. The temptation to dig is always present but must be restrained.

No one may mine in Sri Lanka without obtaining a permit from the State Gem Corporation. This governmental agency regulates all aspects of the trade, from mining to cutting and from land auctions to exports. The agency is extremely conservative and is reluctant to issue permits to mine, even in one's own backyard. Without the permit, digging for gems is strictly illegal.

Rags-to-riches stories of incredible discoveries abound, and occasionally they are true.

A few years ago the world's third-largest sapphire was unearthed; it became the priceless 362-carat Star of Lanka.

In 1984, in the district of Ratnapura, a poor farmer somehow managed to secure a permit to mine on his land. In a single season he found more than $200,000 worth of gemstones, an enormous sum in a country where the average income is about $200 per year. It comes as no surprise that he has permanently given up farming.

In that same year a mineral scout for a well-known mining company was out for a short stroll. While passing a water hole he noticed a woman bathing in a stream and scrubbing herself vigorously with a large blue scrubbing stone. Thinking the man to be a "peeping Tom," she was quite annoyed until she realized that he merely wanted to buy her scrubbing stone. She sold it to him for 2,000 rupees, about $135. She found out later that the man sold it for over $16,000; it was a pure blue sapphire!

The bathing lady, aware that her land was full of this type of "scrubbing" stone, applied herself to the task of obtaining a mining permit and was successful. Today she ranks among the millionaires of Sri Lanka. Undoubtedly she now bathes in any one of her many private bathtubs.

Perhaps it is fortunate that the government imposes rigid restrictions on mining. Sri Lanka is so rich in gemstones that unrestrained mining could result in flooding the world markets. This would reduce the value of such gems as rubies and sapphires far below their absolute value as stones of infinite beauty and finite quantity.

• •

From Out of the Blue

On one bright Sunday afternoon in 1939, a high-powered baseball game was going on in the city park of a midwestern town. The sky above was perfectly clear, with not a cloud in sight, when suddenly a large bolt of lightning came out of the sky and struck the field between the pitcher and first base. The game was halted abruptly, and many of the fans simply fled. Surely this was a sign—from who knows what or where—that something was amiss. But how could this be? Lightning from a clear blue sky? Can lightning strike anywhere at any time?

Throughout the world there are an estimated sixteen million thunderstorms per year. In fact over 18,000 are in progress at any given moment. As is to be expected in any of these storms, thousands of lightning flashes can and do occur. A first-magnitude electrical storm is brilliant in its beauty and majestic in its power and intensity. But, make no mistake about it, lightning is far more than a dazzling display of nature's pyrotechnics. It can be and often is a dangerous killer, far more than people realize. When a hurricane, tornado, or flood strikes, the media give it thorough coverage, and reports of death and damage are quickly flashed across the country and around the world. This is of course understandable, because a bad storm can cause dozens of fatalities and hundreds of injuries. It becomes a memorable incident worthy of the public's attention.

In contrast lightning picks off its victims in ones and twos, which certainly doesn't make national news. Usually it's worth only a few paragraphs in a local newspaper, just as a fatality in an automobile accident would bring attention only locally. Yet consider the horrible cumulative toll of national highway fatalities, which is only a statistic. The destructive power of lightning is well documented, and it certainly is a swift killer of man and beast. The electrical bolts seem almost selective in where they strike and whom they kill or injure. The United States has the dubious distinction of being favored by lightning, for it kills about 500 people annually.

It is extremely dangerous to seek refuge from an electrical storm by running under a tree, yet that is exactly what most people do when caught outdoors during an electrical storm. Trees, being higher than the surrounding terrain, will often initially attract lightning. The bolt may then jump toward a more desirable conductor or run along the ground striking anything in its path. Flying branches and splinters are an additional hazard. Almost one-third of all lightning victims lose their lives because they have sought shelter under a tree; the victims are not always human.

In the spring of 1989 authorities in South Africa reported a particular event concerning two rhinoceroses that were struck simultaneously by lightning. The pair was in the process of mating under a large, isolated tree, unmindful of the violent electrical storm occurring around them. Their lovemaking ended abruptly when a lightning bolt struck the tree that sheltered them, rendering the pair briefly unconscious. They regained their senses at about the same time, staggered to their feet, stared at each other, and abruptly ambled away in opposite directions. They were doubtless blaming each other for the shock and perhaps, in rhino language, were thinking "You're too much for me!"

When the entire planet is considered, the surface of the earth is struck by approximately one hundred bolts of lightning every single second. This may seem like a tremendous number, but over four times as many lightning bolts shoot through the sky without ever reaching the ground.

The length of a lightning bolt varies considerably. Considering cloud-to-earth discharges, the bolt is rarely more than a mile long, although flashes up to four miles long have been reported by authorities. Larger bolts occur when lightning discharges from one part of a cloud to another, and bolts of ten miles in length have been recorded. The longest by far are discharges from one cloud to another. Since distances between clouds may be considerable, horizontal lightning flashes, recorded by radar observations, have been as much as one hundred miles long. It is rare, but it does happen, that one of the long horizontal bolts turns earthward and strikes the ground from what appears to be a clear blue sky. This is what happened during the baseball game described earlier.

Lightning strikes appearing with no apparent source have been recorded throughout history. This undoubtedly has been responsible for the phrase "like a bolt from the blue" or "from out of the blue." The meaning is clear: that something unexpected came from nowhere.

• •

"Riding the Dinosaur"

One of the most informative phases of fossil research is the study of tracks and trails. Ancient prints provide scientists with important data on the physical structure of the animal, its lifestyle, and the environment in which it lived and died.

During the last Ice Age the area now enclosed within the prison yard of the Nevada State Prison in Carson City had been a prehistoric water hole. The numerous well-preserved footprints serve to record the abundance of thirsty animals that came here to drink.

The fossil footprints became quite famous soon after discovery and were not overlooked by early twentieth-century convicts. Those who entered the state prison were routinely greeted by fellow inmates with the remark that, for the next prescribed number of years, they were going to be "riding the dinosaur."

The prints were made nearly 50,000 years ago by a number of animal species, including deer, ground sloth, elephant, and several carnivores such as the wolf and the saber-toothed tiger. One can easily envision the wide berth given the water hole when this great cat came to drink. Since no sign of violence is associated with the cat, it is suggested that other animals were conveniently in short supply as it approached. Thus all it did was drink and wander off.

Several of the prints do record a battle, but this was between two great Ice Age sloths. The rock base, which was then mud, bears the impression of opposing feet, one of which is set deeper at the toes and the other at the heels. This suggests that one sloth was pushing as the other resisted. Close by is a broad depression indicating where a third sloth was squatting, perhaps to watch the fight. Very likely it was the female sloth for whom the males were battling, the winner to enjoy her charms.

Even the wind direction of that long-forgotten day is recorded. On the ridges of each large footprint sand was lodged against the extruded mud. The sand stuck there as the prehistoric wind blew the particles against the upturned ridges, thereby serving as a perpetual record that the wind direction for the latter part of that day was from the southwest, probably in strong but short gusts.

With the expansion of the prison facilities the prison yard was eventually cemented over. Modern prisoners and prison staff as well are generally unaware of the underlying prints, and the phrase "riding the dinosaur" has long been forgotten. Although now hidden from view, this rock record of the past is at least protected from the forces of weathering, erosion, and human vandalism. A few plaster casts of some of the prints are on display in the Nevada State Museum in Carson City.

When the Sahara Was Green

In the heart of the Sahara Desert, in the foothills of the Tassili Mountains, stand several tall, narrow rock pillars of reddish sandstone. One of the pillars is unique: 8,000 to 10,000 years ago a human passed by and carved a picture on it. The composition shows three life-size cattle and a calf drinking at a water hole.

Many other rock carvings and paintings were done in nearby areas during the years that followed. These ancient artworks, combined with discoveries of hippo bones and traces of fishing villages in dried lake beds, clearly indicate that a rather green Sahara existed over 8,000 years ago.

The area was then a broad savanna, populated by gazelles, giraffes, zebras, elephants, and others that naturally inhabit the plains of present-day East Africa. Crocodiles and hippos wallowed in the numerous lakes, and the adjacent mountains, from which many rivers sprang, were covered with forests.

As the Sahara began to dry, the ancient cattlemen were forced to move their herds to the higher elevations of the mountains, which were cool and moist. But the spreading desert was relentless and moved steadily up the slopes, drying out cattle pastures and water holes. The drying process, which went on for centuries, was dramatically recorded by the cattlemen in the rock carvings and paintings of the Tassili Mountains.

More recent paintings record battle scenes between warriors wielding spears and bows and arrows. The men probably fought over the depleting grazing lands and water holes. But in the end the desert was the victor and the people who remained became nomads.

The latest paintings in the Tassili Mountains portray palm trees and camel caravans. These pictures mark the end of a painting tradition thousands of years old.

The fact that the Sahara was covered by savannas and forests as recently as 8,000 years ago has led many people to the false conclusion that the Sahara is only a few thousand years old. However, scientists know from the stratification of ancient rocks that the Sahara Desert originated about 150 million years ago, during the time of the dinosaurs.

But what happened to interrupt the natural aridity of North Africa? To understand this one must go back about two million years in earth history to the last Ice Age. During this cycle of earth refrigeration the ice sheets advanced and retreated several times. The last glacial advance, about 20,000 years ago, reached far to the south. Climatic zones likewise shifted south. With great portions of Europe, Asia, and North America covered by ice sheets, the low-pressure systems, which bring rain, originated in the middle Atlantic rather than in the north Atlantic as they do today. These rain-soaked low-pressure systems crossed the Sahara region instead of northern Europe. The result was a green North Africa, devoid of any deserts.

It was at the end of the last cold phase, about 8,000 years ago, that many of the rock carvings and drawings were created. With the final retreat of the northern ice sheets the world's climate slowly began to return to present conditions. The parching of North Africa was a relatively slow process, and the present extension of the Sahara's southern boundary shows the process is by no means complete.

The Sahara Desert is still in the process of expanding, and it is not alone. Scientists note that the arid lands of the entire world are expanding collectively at the rate of about forty square miles per day, with much help from man. We contribute greatly to the expansion of the world's deserts through deforestation, destruction of the vegetable cover, overgrazing, and generally by consuming rather than replenishing the treasures of the earth.

The Beast of Baluchistan

"There were giants in the earth in those days."—Genesis 6:4

In 1911 Sir Cline Cooper, a British scientist, discovered in what is now West Pakistan the bones of a gigantic fossil mammal. The bones were scanty, leaving him uncertain as to just what the animal had looked like in life. He conjectured that it may have been an unknown form of an extinct rhinoceros. At that time the area was called Baluchistan, so Cooper named his find *Baluchitherium*, or the "Beast of Baluchistan."

A scant four years later a Russian geologist named Borissiak found mammal bones of a similar astounding size in northern Turkestan. Like Cooper, he felt that perhaps they were the remains of an unknown species of rhinoceros. Borissiak did not know of Cooper's find, so he named his mammal *Indricotherium* after the Indrik beast, a legendary Russian monster. According to the legend, the Indrik beast could fly above the clouds, and when it walked or ran on the ground the earth always trembled under its enormous feet. Evidently the Russian believed that the fossil mammal he had unearthed could have made the ground shake when it walked, and he was probably right.

Rumors of this incredible beast were circulated to a scientific expedition to central Asia sponsored by the American Museum of Natural History and led by Roy Chapman Andrews in 1921-24. The main objective of the expedition was to find the bones of man's ancestors, but instead the scientists found dinosaur eggs. Then, on August 5, 1922, in the Tsagn-Nor Basin of Mongolia, members of the expedition found more bones of the Asian giant mammal.

One of the expedition's drivers found a huge bone in the bottom of a gully. Near it was a partial lower jaw with teeth "as big as apples." The size of the bones made everyone speculate that this was the almost legendary Beast of Baluchistan. So it was. The next day they returned to the gully and excavated hundreds of bone fragments, apparently skull parts, but only the front of the skull parts contained teeth. An examination of the teeth and bones by Dr. Walter Granger, chief paleontologist for the expedition, led him to conclude, "I'm sure that the beast is a giant hornless rhinoceros. . . ."

When excavation was complete, they had found and collected the parts of a shattered skull—365 pieces of skull bone all from a single specimen. These were sent back to the museum in New York for cleaning and assembling. Dr. Henry Fairfield Osborn, the museum's chief paleontologist at that time, felt that the bones represented several skulls, but he was proved wrong. For the next three to four months the pieces were fitted together like a jigsaw puzzle. The end result was the first complete skull of *Baluchitherium* ever seen by man. And what a skull! It was four-and-a-half feet long. Even at that size, it was later proved that the giant's head was small for its immense body.

The second find of *Baluchitherium* was made by Liu Hsilkow, a Chinese member of the same expedition. His sharp eyes caught the glint of white bone protruding from the yellow desert sand. Excavation revealed the foot and lower leg of the beast. It was standing upright, as if the beast had carelessly left it behind as it took another step. Fossils are rarely found in this standing position; the only logical explanation was quicksand!

Assuming this, the scientists estimated the right foreleg would be about twelve feet ahead. Carefully measuring the distance, they dug at the predicted location, and uncovered a huge bone resembling the trunk of a fossil tree. Thereafter it was easy to find the two legs of the left side of the beast. The huge animal apparently had wandered into the quicksand bed, sinking to the bottom. It must have been completely submerged in the quicksand, or upon death it would have fallen over on its side. Instead, supported by the encasing sand and its huge weight, it remained upright. A few thousand years ago erosion removed the enclosing rock, revealing the part of the skeleton that had not been eroded. The legs, being unexposed, remained in place. One might say this beast had been standing in the same place for the last twenty million years. Talk about cement shoes!

In 1928, on a later expedition, parts of another skeleton were found. These remains were of a *Baluchitherium* that had died on the bank of a swift river. The flesh had decayed, and the skeleton had fallen apart. Some of the smaller bones had been washed away, but the massive leg bones remained because they were too heavy for the stream to move. Some of the bones were larger than a man's body, the upper foreleg bone being four feet long. A man's forearm bone would look like a mere sliver next to it. The lower foreleg bone, the radius, was over five feet long and so heavy that two men could barely lift it. Truly, this was the leg of a giant, as was later proved to be the case.

When the expedition's finds were completely assembled in New York, the Beast of Baluchistan was ascertained to be the largest fossil mammal known to date. From nose to tail it measured at least thirty-five feet; a mature adult stood about eighteen feet high at the shoulder. When it stretched its neck upward, its nose must have been twenty-five feet above the ground, about nine feet higher than the tallest giraffe. A six-foot man standing under the animal could hardly have touched its stomach.

The *Baluchitherium*, as Dr. Granger suspected, proved to be a giant hornless rhinoceros. A beast of this size had no need for defensive horns such as those possessed by modern rhinos.

The *Baluchitherium* was a browser, eating leaves, twigs, and buds from the treetops. With such high shoulders and so long a neck, it could easily reach into the highest branches. It was basically a peaceful animal. Surely no carnivore would dare attack a healthy adult of such dimensions, but there is no doubt that the herds were stalked by the saber-toothed cats looking for a stray calf or a sick, old adult.

Baluchitherium apparently inhabited only the area of central Asia during the Miocene Epoch, about twenty million years ago. At that time the climate was warm and humid and the land was open plains containing streams, lakes, and lush grass. Although not heavily forested, the area did support a number of trees.

During the middle of the Miocene the collision of the Indian Plate with the Asian Plate caused the Himalayan Mountains to come into existence. These mountains acted like a wall, cutting off the warm moisture-bearing winds and causing the landscape to change. Central Asia dried up, and the trees disappeared. *Baluchitherium* apparently did not migrate, but remained in central Asia despite the change in the basic environment to which it had adjusted. Because the animal had adapted so thoroughly to a warm, moist climate, failure to migrate proved its undoing. Overspecialization has always been one of the chief natural causes of extinction. Thus with the changes brought on by the tectonic uplift of the Himalayas, the highly specialized group of animals could not adapt to the relatively rapid change in environment. And so after being on earth for over ten million years the Beast of Baluchistan became extinct.

The End of Smog

It has been said that if all the automobiles in the world were to be placed bumper to bumper in one continuous line, some fool would still try to pass them. And so it will be until . . . the end of smog.

When the world's petroleum supply is finally depleted, chemists will have to produce gasoline from coal. At modern consumption rates there is still enough coal in the world to last thousands of years. It almost appears that cars will be able to run forever. However, when coal is finally gone, the internal combustion engine at long last will be laid to rest and petrochemical smog will slowly disappear from the cities—forever.

Tarantula Patrol

Although the tarantula is rather fearsome in appearance and size, it is actually quite docile. Its bite, which is quite rare, is less painful and dangerous than a bee sting. If it does bite, the pain comes more from the bite itself than from the venom. Yet humans have an almost unnatural fear of these creatures. This is clearly exemplified by a recent incident in San Francisco.

Having been plagued by a string of robberies, the owner of a jewelry store resorted to desperate measures to avoid bankruptcy. He placed a number of rather large live tarantulas in his showcase window facing the street. He also put a sign in the window with a warning for all to see: "Danger! This area is patrolled by tarantulas."

Not surprisingly, the robberies ceased almost immediately.

• •

Killer Mice

During the 1950s Hollywood ran wild in its science fiction renderings of giant insects, giant reptiles, and other huge creatures. (The ultimate may have been attained when a film was made about "killer tomatoes.") But reality can be stranger than fiction, and killer mice are a good example.

The name of these rodents was not derived from a bad horror movie, but from their very real nature. There are three species of killer mice, *Onychomys* sp., often called *grasshopper mice* since grasshoppers are their preferred food. *Onychomys* are fairly common and widespread in the central and western United States. They look like ordinary field mice—brownish, small, with big ears and long whiskers. Taxonomically they have been classified as murids, which makes them members of a common family of normally vegetarian mice. What they do, however, is anything but normal or vegetarian.

The grasshopper mice are ferocious predators—in miniature of course.

They hunt by night in large packs, seizing their hapless prey and tearing it limb from limb in a matter of seconds. Nor does the carnage stop there. When the prey is abundant, they raise their pink-tipped noses in the air and howl like tiny wolves—apparently calling their friends and relatives to join the feast.

As their name implies, these killer mice are most partial to a grasshopper dinner, but a furious appetite added to a lust for killing causes more than insects to become their prey. They will not hesitate to launch a mass attack on field mice, lizards, or any other small animal that blunders into the path of the hunting pack!

Not surprisingly, farmers regard the grasshopper mouse—the scourge of grasshoppers, cicadas, caterpillars, moths, beetles, and small rodents—as a welcome intruder. With 95 percent of its diet consisting of insect or animal food, the grasshopper mouse not only spares crops, but also eliminates crop destroyers with incredible efficiency. So the killer mice are to be doubly praised.

• •

A Sword from Heaven

Because iron from meteorites is very durable and malleable, man has been quick to take advantage of these properties. Apparently, meteoritic iron has been used in the forging of weapons for thousands of years.

One of the earliest discovered is a dagger that was unearthed by archaeologists several years ago at Ur of the Chaldees. The dagger, made of meteoritic iron, had been forged at least 5,000 years ago.

Weapons made from such heavenly iron have been found all over the world, from Mongolia to Mexico, from Europe to Africa, and even in the United States. Perhaps the most famous American knife is that of Jim Bowie, which was reputed to have been made from meteoritic iron. His legendary knife was lost as "booty" to an unknown Mexican soldier following the fight at the Alamo in which Bowie lost his life.

The sword of Attila the Hun shed much human blood, as recorded history attests. He called the weapon his "sword from heaven," a most appropriate name since it was fashioned from meteoritic iron. Apparently he had seen the meteorite fall and somehow managed to recover it. In his megalomaniacal mind it must have appeared as a special gift from the gods to assist him in his war against mankind. And so he forged it into a sword; the rest is deadly history.

• •

Moby Dick Was Real

The sperm whale, largest of the toothed whales, about seventy feet in length and weighing fifty tons, can and does at times display tremendous power and ferocity.

Titanic battles often occur between males, usually for female harems. They ram each other with their tremendous heads and slash at each other's flukes and flippers with their fearsome eight-inch teeth. It is not unusual for combat to end in death for one of the participants.

Lone bull sperm whales caused spectacular maritime disasters during the nineteenth-century whaling era. The sailors in small whaleboats, as often as not, never even got to harpoon the whale. The enraged bull would bear down on them, crunching the small boats to splinters in its jaws. Or it might choose to smash the boat to splintered wood with a tremendous whack from its fluke.

As if this were not enough, at times the enraged bull would turn and batter the mother vessel itself. During the nineteenth century a number of large wooden whaling ships were sent to the bottom when a bull sperm whale used its enormous head to punch massive holes in the ship.

This type of disaster was usually accomplished by large rogue bulls who became so well known that they were even given names. The most notorious was a large white sperm whale known as Mocha Dick. He was the most feared of rogue whales and promptly became the terror of the whaling waters. He destroyed a number of mother ships through unprovoked ramming. The huge holes in the ships' sides quickly sent them to the bottom, often with their crews.

In 1842 Mocha Dick rammed a coastal freighter off Japan and turned on three whaling ships that came after him, ramming and chewing up small boats and men alike. His last victim appears to have been the *Ann Alexander* in 1851. The whale first demolished the small boats containing the harpoon crews and then turned on the large whaling ship. The tremendous head easily bashed a large hole in the side, causing the ship to sink into the sea.

The eventual fate of Mocha Dick is unknown. Perhaps he was killed by another rogue or just died of old age, but whatever his end, his disappearance from the sea was a great relief to the whalers of the era.

When Herman Melville wrote his classic tale *Moby Dick*, the basis for this story of a great white whale was, indeed, Mocha Dick.

• •

What Did the Phoenicians Discover?

The use of window glass is considered by most people to be modern. Glass windows, however, have been found in the ruins of Pompeii. The rarity of such windows implies that glass must have been a luxury used only in the finer homes and buildings of that time.

Glass is manufactured from silica, one of the most important and widespread elements in the earth's crust. In its mineral form silica is called *quartz*. A typical sandstone is usually composed of tiny grains of quartz held together by some material acting as a cement; often the cementing agent is also silica.

Sand particles are formed from the weathering and erosion of a parent rock, usually igneous (rocks formed from a molten state). Logically, then, in its early stages the sand will have particles of many other kinds of minerals mixed with the quartz grains, depending on the mineralogic content of the parent rock. Sand grains that lie along shores are usually subjected to the endless wash of waves and often become remarkably pure in quartz content. The water dissolves or carries away the softer minerals, leaving behind the harder quartz grains. Sandstone made solely of quartz grains is used regularly in the manufacture of glass.

How the glass-making process originated has been lost in antiquity, but the Roman naturalist Pliny the Elder stated that glass was discovered accidentally by some Phoenician merchants.

Pliny appears to have been a great admirer of the Phoenicians, crediting them with many discoveries, including the invention of trade. Doubtless he exaggerated, but scholars do recognize that they were the first traveling salesmen.

The Phoenicians traveled along the northern rim of the Mediterranean Sea, establishing colonies, trading posts, harbors for their ships, and sheds for their goods. They made many short-haul commercial trips, usually sailing by day and always within sight of land. The Phoenicians not only traded their own products but also trafficked in tin from Spain, copper ore from Cyprus, and whatever else was available and considered valuable. Their travels also included the coastal area of West Africa. Upon landing they spread their wares on the beach, whereupon they returned to their boats and raised a signal. Natives came out and laid gold beside the Phoenician goods as a barter price, after which they retired. The mutual comings and goings went on until both sides were satisfied. The Phoenicians would then sail away with the traded goods. Both Pliny and Herodotus reported that each dealer felt satisfied that he had not been cheated.

What particularly impressed the ancient world about the Phoenicians was their skill at long, daring sea trips. During the seventh century B.C., the Egyptian pharaoh Necho, searching for a viable connection between the Red Sea and the Mediterranean and having no notion of the size of Africa, sent some Phoenicians south of the Red Sea. For the next three years they circumnavigated Africa—a feat that would not be repeated for nearly 2,000 years, until the time of Vasco da Gama.

The Phoenicians did stop each year at seed time to plant and harvest; then they continued their journey. When forced to sail at night, they navigated by a bright star in the constellation Ursa Minor. Then known as the "Phoenician Star," it is now called Polaris or, more commonly, the North Star. The intrepid explorers did not find a passage to the Mediterranean, simply because there wasn't one. A passage was created in 1869, currently called the Suez Canal.

According to Pliny, on a long-forgotten evening after landing on the coast of Palestine near the mouth of the Belus River, the Phoenicians set to the task of preparing their evening meal. Being unable to find proper rocks on which to set their pots, they obtained some cakes of saltpeter from their ship's cargo and placed their cooking vessels on them before lighting a fire. The heat from the flames caused the saltpeter and quartz sand on the shore to melt. These combined into streams of an unknown fluid, which hardened to a translucent substance later known as glass!

• •

Kapiolani

Pele, the mythical goddess of volcanoes, held a rather stern grip on the superstitious minds of early Hawaiians—that is, until the time of a woman named . . . Kapiolani.

In the early 1700s, a group of Hawaiian soldiers returning from a victorious battle stopped at Kilauea, home of Pele, goddess of volcanoes, to pay homage. As they prayed to Pele, thanking her for their victory and safe return, the volcano erupted without warning and killed about one-third of the men, mostly with poison gases. Footprints of the stricken soldiers are well preserved in volcanic ash along the old trail in the national park. In the minds of the survivors, Pele, for no apparent reason, was angry and punishing them.

Intense belief in Pele held sway until 1824 when, three years after she had been converted to Christianity, Kapiolani, wife of the high chief of the Kona District, resolved to show her people that Pele was nothing more than a heathen superstition. Priests, husband, and friends all prophesied a fiery doom for her.

She traveled well over a hundred miles across rugged terrain to Kilauea, where Pele was reputed to dwell. Her prestige and influence preceded her, and during the journey increasing numbers of people followed her to the volcano, staying at a respectable distance lest the fire goddess misunderstand why they were following her.

Reaching the fire pit of Pele, Kapiolani immediately broke taboos by eating sacred ohelo berries without offering any first to Pele. The gathering crowds of people gasped as she threw rocks into the sacred crater. She then descended several hundred feet to the very edge of the lava lake and openly defied Pele to destroy her.

Remaining at the edge long enough to prove her point, she then ascended the crater. When she returned unharmed, the people began to realize that Pele *was* just a superstition. Many soon converted to Christianity, a practice that gradually took hold and spread throughout Hawaii. Among the converts was the high priest of the volcano!

Alfred, Lord Tennyson later paid homage to this very brave woman by immortalizing her in a poem he simply called "Kapiolani."

Some Hawaiians do still believe in Pele, or so it appears. Periodically they gather at the top of Kilauea and throw offerings into the crater. Quite likely this is done to impress the tourists, because the offerings seem to be made only in their presence. However, one worshiper, in recognition of Pele's unusual drinking habits, was observed to throw an unopened bottle of gin into the crater.

In recent years Pele, if she exists, seems to have been celebrating the seventy-fifth anniversary of the nearby Hawaiian Volcano Observatory and the opening of its new facilities. In the few months since its opening Kilauea has added eighteen acres of lava to the island of Hawaii.

The current eruption actually started in 1986, and as of this writing the volcano is still in an eruptive state. From 1986 to 1989 it has produced a record-breaking 850 million cubic yards of lava, enough to cover, to a depth of over thirty feet, four lanes of an interstate highway from New York to San Francisco!

• •

The First Family

It was during August of 1964 that a team of government geologists encountered one of the furies of nature that could have been a prelude to disaster. The senior geologist, recognizing the inexperience of his two assistants in dealing with the hazards of the Arizona desert, took special pains to educate them. He recognized, however, that some things are learned most effectively through experience.

The mission of the group was to map some of the unexplored canyons. Late one morning they were driving their jeep in a narrow V-shaped gully. The senior geologist stopped the vehicle and looked toward the east, where he observed a violent cloudburst in progress, while overhead the sky was clear and blue and the sun was shining. His warning of a possible flash flood in this gully fell on deaf ears, so he herded his crew into the jeep and drove down the streambed. He then stopped abruptly; although the storm in the east had ended, he heard sounds that resembled the echo of thunder, and they were growing louder and louder.

Realizing what he was hearing, he yelled to the men to abandon the jeep and climb out of the gully. The men just sat and stared at their leader, doubtless convinced that he had been in the desert too long and had finally flipped. But they were becoming apprehensive, for the roaring sounds were getting nearer and louder.

The senior geologist leaped out of the jeep and climbed the steep gully walls. The two assistants, having no better plan of action, followed him somewhat blindly. They had barely reached the top when, around a bend in the gully, a wall of water at least fifteen feet high erupted into view. In an instant the jeep was gone. It was found the next day nearly a mile away, a complete casualty.

Desert people worldwide avoid arroyos, or narrow valleys, as campsites. In fact many a bedouin has a well-defined phobia about being caught in a deep gorge. They have learned through bitter experience that stony landscapes with little vegetation will concentrate rainwater in ravines and gullies. When this happens, a deep ditch can become a death trap, with a flash flood hitting even hours after the rain has stopped and miles away from the original downpour. In such floods the water can rise ten or twenty feet in a matter of a few minutes, drowning everything in its path.

Recently in the northern Sahara four tourists, along with their Arab guide, observed a violent cloudburst occurring about twenty miles east of them. Their point of observation was, of course, from a deep narrow ravine. Their guide became agitated, and when he heard thunderous noises coming from a distance, he ordered the group to abandon their vehicles and climb out of the gully. Naturally they refused, thinking he had been in the sun too long. But as the thunder got louder and closer, he drew his pistol and ordered them to climb the steep walls, emphasizing his command by firing the gun at the ground in front of their feet. They barely reached the top when a huge wall of water swept into view and carried away their vehicles and abandoned supplies.

The fee the grateful people paid their Arab guide at the conclusion of their tour was far above the amount originally agreed on. They realized that the guide, by resisting their objections, had saved their lives!

The threat of flash floods in the deep ravines of arid lands has always been a menace to man. Such was the fate of the first family. They were involved in an incident not unlike those described above in modern times, but they were to pay with their lives.

Between 1973 and 1977 Dr. Donald C. Johanson, a leading paleoanthropologist, was codirector of several consecutive international expeditions in search of early man in the Afar region of northeastern Ethiopia. About 250 hominid (early man) fossils were found in the ravines and tributary valleys of the Hadar River. More fossils would doubtless have been found if expeditions could have continued, but internal strife in Ethiopia curtailed further exploration indefinitely.

Radiometric dating has shown that the Afar fossils are well over three million years old. They represent an extraordinary variety of teeth and bones of about sixty-five individuals; the most famous is called "Lucy."

She was found during the 1974 expedition when, on a November afternoon while exploring with a colleague, Johanson noticed a fragment of an arm bone poking from a slope. A three-week intensive sieving job on the slope uncovered the remains of the oldest known hominid. About 40 percent of an entire skeleton was recovered—a new high in paleoanthropology because hominid remains from a single individual this ancient have never been found in such abundance. The pelvic bones showed the specimen was female, and erupted wisdom teeth suggested she was about twenty years old when she died. Her height, estimated from the thigh bone, was about three-and-a-half feet. Later finds of this hominid species indicate some males may have reached four feet in height. Although these creatures generally measured less than four feet and had a relatively tiny brain for a hominid, they all walked erect. From this it can be inferred that man achieved bipedalism before time and genetics had authorized the enlargement of the brain.

When this early hominid was first discovered, the realization of its importance set the camp into intense excitement. Nobody slept that night; the scientists talked and the beer flowed. The camp tape recorder incessantly played the Beatles' song "Lucy in the Sky with Diamonds." As the song was played at full volume over and over again, it seemed inevitable that sometime during the night this most ancient female would become known as "Lucy," and so Lucy she was and still is.

She was later assigned the scientific name of *Australopithecus afarensis*, a name steeped in controversy because some scientists do not believe Lucy represents a new species.

Now the stage is set for the first family. In the fall of 1975 another startling find was made in the same area—a slope virtually littered with the bones of some of Lucy's relatives. The scientists spent weeks on intensive excavation of that particular slope, systematically combing the entire hillside. Tons of the gravel surface were carried down the slope to be sifted, load by load, through coarse sieves. Ultimately the hillside yielded over 200 teeth and pieces of bone. It was impossible to reconstruct the skeletons because of the way the remains were jumbled and scattered down the slope, but all agreed duplication of specific parts made it clear that at least thirteen individuals were represented, including men, women, and four children.

These hominid fossils came from a common source in the stratigraphic horizon near the top of the slope. Those found on the slope had been washed out through erosion and scattered on the incline. The source horizon still contained nearly twenty more fossil pieces, all at the same stratigraphic level and obviously still in place. The implication was clear: before erosion had scattered their remains, all had lain in close proximity and therefore presumably had died together.

But what killed them—an epidemic, a fight, perhaps an accident? No, because had this been the case, the remains would have been gnawed by scavengers or the bones cracked open for their marrow. None of this was evident.

An examination of the horizon that contained the remains gave the answer. The relative thinness of the stratigraphic layer, which consisted almost entirely of fine clay, suggested a single event, such as a sudden flood. However, one must consider that a sedimentary deposit originating from rushing water would have carried a mixture of sediments larger than clay size. Evidently this is not where they died, but where they came to their final rest.

Everything—stones, mud, and sand—must have been carried down in a rush to a lake edge or some other more level open area where the velocity of the water was controlled. The bodies were immediately deposited, possibly in a heap. As the remainder of the floodwater flowed and ebbed, the fine material it was carrying settled over the bodies and buried them from sight for the next 3.4 million years!

Since they evidently lived together, they surely must have been members of the same family, so Johanson dubbed them "The First Family."

Now let's go back to the time of Lucy. Nearly three-and-a-half million years ago a small group of very primitive hominids were encamped in a deep ravine, probably at the start of the rainy season. Some were sleeping, some were feeding, and others were grooming their neighbors. They were aware of an intense cloudburst occurring in the distance, followed by thunderous noises that kept getting louder and louder.

It is a common belief that humans living close to nature are very wise in the ways of the wild. This is doubtless true for modern *Homo sapiens,* but during the time of the first family hominid reasoning was relatively embryonic. So, unable to comprehend what the noises implied, they doubtless shifted around uneasily, trying to decide what to do. They all turned and stared in disbelief and horror as a wall of water swept into view. The waters quickly inundated the entire group, and they became part of the rushing torrent. Mercifully they all drowned almost immediately, and their agony was ended.

As the water flowed out of the confinement of the deep gully, it spread out on the open plains and entered a lake. The bodies were dropped in close proximity as the water's velocity decreased. The flowing muddy water, representing the tail end of the flood, quickly covered the bodies with fine-grained silt and mud, and this impromptu grave protected them from scavenging animals. Their ordeal for survival in a hostile world was over, and they were at peace.

As the wise ones have often said, "Families that play together stay together"—in this case forever.

• •

CHAPTER EIGHT

The Return of a Native

After sixty million years of evolution in North America, the camel crossed into Asia via the Bering land bridge of the last Ice Age and became extinct in its homeland. When the United States Army reintroduced them into America, who can deny that for the camels it was . . . the return of a native?

In the mid-nineteenth century, at the urging of Jefferson Davis, then secretary of war, Congress appropriated $30,000 for importation of camels to be used as beasts of burden in the arid American Southwest. In 1856 some thirty-four camels and a number of Arab drovers were imported into this country. The voyage across the sea was long and, for the Arabs, who had never been aboard a ship before, torturous. They became so seasick that upon arrival in the United States all but two deserted. The fate of the deserters is unknown. It is doubtful that they would ever undergo life aboard a ship again, even to return to their homeland.

The two Arabs who did not desert became part of the army and were issued regular army rations. This led to an embarrassing situation. While on a short journey the army officer in charge noticed the camels swayed, zigzagged, and were generally unsteady. Something was definitely wrong with their gait. Upon investigation he quickly found the cause of the camels' trouble. They were drunk!

The Arabs, being complete teetotalers, didn't know what to do with their daily beer ration, so they poured it into the camels' water pails. The army handlers who observed this practice thought it was part of the camels' regular menu, so they supplied the desert animals with large amounts of beer. The camels liked the new beverage, lapped it up, and staggered around for the remainder of the day. The inquiring officer quickly ended these antics.

In time camels were added to the new United States Camel Corps. The animals were used as beasts of burden, carrying army supplies, particularly

in the deserts of California and Arizona. Regular caravans of camels from Fort Tejon entered the city of Los Angeles, plodding down the dusty streets to the heart of the city, where the army maintained a large camel stable. This was also headquarters for the Camel Corps in California, and today it is the site of the *Los Angeles Times* building complex.

Despite the many claims of success for the Camel Corps, the project was doomed. Inexperienced army drivers could not cope with the camels' short temper, and the sight of these strange beasts caused numerous stampedes among cattle and horses. While their usefulness was beyond doubt, it was seriously limited by their nuisance factors.

The project was finally abandoned with the onset of the Civil War. Many of the camels were sold to circuses or to private owners. Herding of the camels still in army service became very lax, and many of the imported animals escaped into the desert and ran wild. They had no trouble adapting to their ancestral homeland, and they thrived, especially along the Gila River in Arizona. In the years to come they were frequently rounded up near the river and sold to traveling shows.

Indians soon discovered that camel meat was a tasty and hearty dish and hunted camels like buffalo. With the continuous onslaught their numbers rapidly decreased. The last camel hunt took place in 1899 near Yuma.

In the years that followed there were persistent rumors of camel sightings; even now rumors surface occasionally, but they are becoming relatively rare and none are substantiated. However, as late as 1941 fresh camel tracks were definitely seen in the sands near Douglas, Arizona. One cannot help speculating that in the untraveled wilds of the Sonoran Desert of Arizona and adjacent Mexico a few roving wild camels may still exist.

Just before the turn of the century an incident occurred that enriched the lore of the American camel. It had the makings of a Hitchcock thriller, the major difference being that this story is true.

A rancher's wife living on Eagle Creek, Arizona, was stomped to death by what witnesses described as a "Red Demon with a man on its back." Just two nights later the same red demon wrecked a prospector's campsite on Chase's Creek. The man, frightened out of his wits, ran away, screaming incoherently about the wild demon. During the weeks that followed reports continued to come in from other eyewitnesses who had encountered the red ghost.

The mystery was solved in February 1893, when a man named Mizoo Hastings spotted a large intruder in his turnip patch. He pushed a buffalo gun out of a window in his ranch house and fired. The career of the red demon was brought to a sudden and abrupt end as the creature fell dead.

The red demon was discovered to be a huge camel that had a human skeleton tied on its back with leather thongs. The human had been bound so tightly that the thongs cut into the camel's hide. The constant pain that resulted must have deranged the unfortunate animal, probably accounting for its hostile behavior.

Only then did the old-timers recall a story passed among early settlers in Arizona. At last the story was believed. A few years before, the Navajo Indians had killed a Mexican shepherd, lashed his body tightly to the back of a wild camel, and then chased the animal far out into the desert, to warn intruders to stay out of their land. Unfortunately it didn't work!

• •

Extraterrestrial Trauma

Throughout history man has lived in superstitious fear of celestial events. On occasion, however, the results have been favorable. A total eclipse that occurred in 585 B.C. ended a six-year war in Greece. A battle was actually in progress when the solar event turned day into night. The conflict was immediately terminated as the soldiers hid their faces behind their shields in paralyzing terror. They believed this was a warning from the gods to cease fighting, and a truce was signed that evening.

Ancient people had widely varying methods for dealing with an eclipse. These ranged from firing arrows at the darkening solar disk to sacrificing a virgin to the angry gods. Whatever the method used, the sun was always "rescued" and returned to its natural brilliance.

Modern soldiers deal with an eclipse in a more effective manner than did the ancients. During the Vietnam War, Lon Nol's Cambodian troops halted their campaign to fire automatic rifles at the darkening sun. They succeeded in rescuing the sun, of course; the darkness moved on, and light was restored. Having won the "battle," the soldiers returned to the skirmish close at hand, which many discovered to be far more deadly.

In 1973 teams of scientists moved into central Africa, where a total eclipse could be observed best. When they arrived to study the celestial event, they were met with determined hostility from the natives. These people believed that the scientists were trying to kill the sun and were therefore responsible for the darkening solar disk. The hasty retreat executed by the scientific team was anything but dignified.

Fear of celestial events is not restricted to eclipses. Although they are less spectacular, comets are also enshrouded in awe and mystery. Bennett's Comet in 1970 caused much concern among Arabs, who feared it was an Israeli secret weapon!

Because comets appeal to the superstitious nature of mankind, the forecasters of doom have been quick to capitalize on the plight of the gullible. The Incas regarded comets as intimations of wrath from their sun-god, Inti. The high priests, not surprisingly, encouraged this belief and often insisted on an additional human sacrifice to appease the angered god. Apparently this action always enhanced the integrity of the high priests.

When Halley's Comet passed near the earth in 1910, the prophets of misfortune were out in force, spreading their tales and creating panic. Several people tried to beat doomsday by committing suicide. In Oklahoma a fanatical clan that called itself Select Followers was actually preparing a human sacrifice—a virgin, of course—when the police, acting on a tip, arrived in time to stop the tragedy!

Not everyone reacted in panic as Halley's Comet approached; in fact one enterprising American manufacturer in San Francisco took advantage of the doomsday situation. He made and sold comet pills! Just what the pills were supposed to do everyone conveniently forgot. Nevertheless the man made a considerable sum of money, proving again that P. T. Barnum was on target when he announced, "There's a sucker born every minute."

The world did not come to an end, and the comet went on its way. Interestingly the tail of the comet did actually sweep the earth with no noticeable effect.

Prophets of doom have been on the scene for almost every predictable cataclysm. There was a near panic in 1524 when a group of London astrologers predicted that on February 1 a great tide in the Thames River would wash away 10,000 houses. During the month of January well over 20,000 people left the city and fled to higher ground. Normal tides flowed and ebbed, but no great flood occurred.

Naturally this forced the doomsday prognosticators to go back to the drawing board. In reconsidering the records the astrologers found they had made a slight error. The catastrophe would occur one hundred years later, in 1624 rather than 1524. So the unsuspecting believers could rest easily, knowing that doomsday would not come during their lifetime.

Even today the prophets of disaster periodically frighten the public with predictions of cataclysmic events. When the forecast disaster does not happen as scheduled, they often cover their miscalculation by simply saying nothing or, quite often by coming up with an alternate prediction that believers will accept, almost without reservation.

• •

King for a Day

A most gratifying theme for fiction is the story of an insignificant person being elevated to the position of royalty for a brief period, usually just for one day.

Ironically the event is based on an actual custom that was widespread in ancient Babylonia. On New Year's Day it was customary for the reigning king to appoint a person of low stature to be "king for a day." The unfortunate appointee would be wined and dined and given anything he wanted. He was appropriately designated "unfortunate" because the next day he would be put to death! On one New Year's Day, however, the custom backfired. King Enlil-Imitti appointed his gardener to the post, and during the celebration the real king died. He appears to have drunk too much wine, and quite probably he received a little help toward his demise. And so Enlil-Bani remained on the throne and ruled for the next twenty-four years. It was noted that he never appointed a gardener to the position of "king for a day."

In contrast with such short reigns, the longest in recorded history was held by Pepi II, who succeeded to the throne of Egypt in 2272 B.C. and ruled for at least the next ninety years. He was doubtless a mere infant when he ascended the throne, probably not even old enough to sit on it. His feet certainly did not touch the floor when he was first placed on the royal seat, but they were firmly implanted for the decades to come. Records clearly show that during his adult years Egypt prospered under his rule but fell apart a mere two years after his death.

• •

Screams in the Night

Since human beings do not see well at night, nocturnal animals are those with which man has the least experience. It is a fact that when one of man's senses is impaired the others become much sharper. By wearing a blindfold for two or three hours a person finds that his hearing becomes much more acute. Campers in the forest at night see very little, but their hearing often gives rise to flights of imagination.

During World War II soldiers standing guard at night magnified each sound they heard, convinced that enemy soldiers were sneaking through the bushes. Occasionally they would respond by opening fire, much to the consternation of the men who were seeking some solace in sleep. The lurking enemy was usually falling leaves.

In North America screams in the night have led to frightful tales that were stretched considerably by imagination. In the early days of American history European settlers moving west were often awakened by a strange and terrifying sound in the dead of night. The scream sounded so horrible that the bravest of men crouched closer to the fire and attributed it to the devil. It was often described as the cry of a woman being strangled or of someone extremely frightened or in terrible pain.

The Indians believed the scream was the sound made by men of magic who transformed themselves into hideous monsters. These monsters were able to run through the forest killing people with nothing more than their frightful screams. For over a century this sound was reported over and over again. People who never entered the forest at night scorned these reports, referring to them as "old woodsmen's tales."

Nobody seems to know exactly how, when, or by whom the legend of screams was solved. Scientists now know the screams, which resounded through the forests at night, are the cries of the female mountain lion calling to her mate. Although the mountain lion is quite capable of growling, snarling, and hissing, it usually projects a small voice and makes sounds akin to the chirp of a canary.

But when the female cat is ready to breed, she makes no secret of it and proclaims it loudly to every creature within hearing distance. Since mountain lions are solitary animals, the female's cry must carry far. As for her mating scream, the only real way to describe it is that it sounds terrifying to humans. To the male mountain lion, however, it is a beautiful and inviting sound, and he will follow it to its source.

In a relatively short time the population of the forest will increase by the birth of several kittens. The mating call has served its purpose, and the species lives on.

• •

Mavericks of the Solar System

In certain regions of our planet strange glasslike pieces of rock have been found that have proved to be quite a puzzle to geologists. They are brown to green in color and are shaped like rods, spheres, disks, or buttons. They are generally smaller than two or three centimeters in diameter, with a few being over ten centimeters.

Usually found in clusters, they have been discovered in fewer than a dozen locations throughout the earth, mainly the Ivory Coast, Indochina, Australia, Texas, and Czechoslovakia. Scientists refer to them as *tektites*. All tektites found in the same area are similar in structure and composition and have the same geologic age. All have the appearance of having once been partially molten. Perhaps one of the strangest things about them is that they have no geologic connection with the terrain in which they are found.

There are many theories as to the origin of tektites, but most known facts seem to point to an extraterrestrial origin. From which part of space they came and why they are so different from meteorites are more elusive questions.

Increased data derived from research indicate a possible lunar origin, a theory that is becoming more acceptable within the scientific community. Tektites appear to be fragments ejected into space from the moon as a result of heavy meteoric impact.

Given the low escape velocity from the moon (about one-and-a-half miles per second) following impact by a large meteorite, fragments of the lunar surface can be expelled into space with sufficient velocity to avoid recapture by the moon. Some of these fragments can be thrown into trajectories that reach the earth, where they go into orbit. The ejected material is fused on impact by the meteorite but cools on its journey to earth. When its earth orbit decays, it fuses again because of friction with the atmosphere. The results are the small glasslike rock fragments of specialized shapes strewn in very specific localities.

If all this is true, then each group of tektites would result from the formation of a large lunar crater by the impact of a specific meteorite. This would explain why tektites of one locale are the same age and of similar composition and why they differ from those of other regions.

Recent research on some Australian tektites shows they are remarkably similar in composition to those rocks found at the edge of the lunar crater Tycho, which was examined by the U.S. probe *Surveyor 7* in the landing of January 1968. Extensive computer studies on possible trajectories of the Australian tektites indicate that they could have been formed by a shower of particles originating from Tycho.

Further evidence seems to support this. Since the Australian tektites are about 700,000 years old, it follows that the great lunar crater known as

Tycho was created only 700,000 years ago by the impact of a gigantic meteorite.

Perhaps sometime in the near future the moon will again collide with a large meteorite and a new crater will be formed. When this happens, the earth may once more be hit by a shower of fiery lunar drops, and another field of space mavericks will be sown.

• •

Streets Paved with Silver

During the early days of mineral exploration in the United States one of the greatest silver-producing districts was Virginia City, Nevada. It did not begin that way, however, for prior to 1865 Virginia City was a gold-mining district. The abundant blue-green rock was ignored as a potential ore and was used extensively to pave the city streets.

Visiting Mexican prospectors, who had worked the silver mines in Mexico, immediately recognized the rock as silver ore. Quickly, assays showed the blue-green rock to be extraordinarily rich in silver. The streets of Virginia City were completely torn up by the local citizens within thirty-six hours! Shortly after the streets were "mined," serious shaft mining began.

For the next few years the district produced so much silver that the world markets were flooded, causing several countries to go off the silver standard. Yet with all this intensive mining done in the past, probably 90 percent of the silver ore still remains in the ground under Virginia City, untouched!

• •

Garlic

Garlic, which suggests good eating in many ancient languages, has been part of the human diet for thousands of years. As early as 3000 B.C. Chinese scholars sang its praise. The Great Pyramid of Giza was built by garlic- and onion-munching slaves. The labor in the pyramid construction may have even been tolerable as long as the slaves stuck to their work and avoided breathing on each other.

No responsible Phoenician or Viking ever embarked on a long voyage without stuffing his sea chest with garlic. In addition to being a staple in ancient diets, garlic has always been immersed in superstition, both good and evil. Since antiquity people have worn strings of garlic around their necks, carried it in their pockets, or hung it over their doorways and windows. The purpose was to ward off everything from vampires and werewolves to the plague. This custom is still practiced in certain regions of Europe.

One of the strangest superstitions involving garlic was in force during the Renaissance, when it was believed to be a very potent love charm. Those readers familiar with Boccaccio's *Decameron* will recall the tale of a love-stricken young man who sent his chosen lady love a string of fresh garlic to win her affection.

Belief in its love potency was so firm that all a man had to do was prevail on a clergyman to bless the garlic. This undoubtedly clinched its potency, because if he then presented the garlic to the lady of his choice she would be unable to resist him—or was it the garlic? (This practice doubtless took place before the invention of kissing!)

At times toreadors have worn wreaths of garlic around their necks to protect them from being gored. Its effectiveness doubtless depended on whether the bull was repelled by the smell. A guarantee of a good race was to rub garlic on the horse's hoofs, and a dream of possessing garlic would certainly mean that good fortune was on its way. On the other hand, to dream of giving away garlic was a sure sign of forthcoming ill fortune for the dreamer.

Perhaps the most outstanding feature of garlic is that it has been a perennial staple in folk medicine, hailed as a panacea that cures many ailments. This sentiment still prevails, adding headaches for physicians who are consulted only after the garlic treatment fails. The irony of its preventive-therapeutic reputation is that modern research indicates that some of the medicinal claims for garlic are valid. Garlic contains the amino acid alanine, which has antibiotic and bactericidal effects. Further studies have shown it may also help prevent the development of arteriosclerosis (hardening of the arteries).

Soviet doctors are currently studying the effect of garlic on certain types of cancer, while in Japan research is being conducted to determine its effect on lumbago and arthritis.

The onion, a close cousin of garlic, apparently also has some definite medicinal properties. Scientists have recently found that raw onion contains quantities of a substance called prostaglandin A, which is known to be effective in lowering blood pressure. Although results are not conclusive, injections of prostaglandin A in rats have been remarkable in regulating circulation. Since the circulatory systems of rats and humans are very similar, the same effects may be expected on man.

So if your friends don't object, indulge in a succulent raw onion for good health and good eating.

• •

Cheetah!

The cheetah, *Acinonyx jubatus*, is the most uncatlike of all cats. With its nonretractable claws, a unique feature among cats, and its long, straight legs, it is a rather doggy feline.

Unquestionably the cheetah is the fastest four-legged creature living today. In just two seconds it can accelerate from a standing position to forty-five miles per hour. When in full stride it can cover 103 feet in a second—seventy miles per hour! And many claim the animal can go even faster.

It is this remarkable speed that makes the cheetah a most successful hunter, for it merely runs its prey down. The speed factor, however, does limit the cheetah to open savannas. One can hardly expect a runner to get up a good head of steam in wooded country.

The favorite prey of the cheetah is the impala or Thompson's gazelle, either of which usually weighs less than the cheetah. When the cat charges, the chances are the cheetah will pass the prey by if it stands still. Impalas often do just that, either while they contemplate what to do next

or, more likely, because they are paralyzed by fear. The cheetah must pursue its prey in flight because of its pattern of aggressive hunting.

As the cheetah pulls up beside the fleeing impala, it will swing out a paw in midstride and trip the hapless prey. This is quite a remarkable feat considering that the cat is traveling at a clip of over one hundred feet during the second it takes to swing a paw. When the impala goes head over horn, the cheetah slams on its brakes and pounces.

For all of its speed, the cheetah is not dangerous to man. Unprovoked attacks are, as far as the records indicate, unknown. Even provoked attacks are extremely rare and are seldom, if ever, fatal.

Man is simply ignored by the cheetah, as was evidenced by an incident in East Africa. A group of tourists stopped in a Land Rover to photograph a gazelle that was being stalked by three cheetahs. They were able to get many more unusual photographs than anticipated because the cats actually used the car as a blind, slipping into position by the front wheel to launch their attack. The cats were so close that the men could have reached out and petted the animals, which completely and thoroughly ignored them. Needless to say, the hunt was a success.

Since most everyone will agree that the cheetah is the fastest creature on four legs and not dangerous to man, why is it not used in the sport of racing? Well, it has been tried.

In 1937 eight trained cheetahs were imported to England and were matched against greyhounds on a well-known racetrack. The cheetahs, although much faster than the dogs, absolutely refused to cooperate and simply would not run. Apparently they saw no purpose to it. One yawned noticeably and just lay down, and the others followed suit.

Another race was tried, with cheetahs competing against each other. It worked at first. Away they sped, but the race was short: if one pulled ahead, the other would simply stop. One cheetah running on its own covered half the track at a breakneck speed, too. But suddenly it stopped abruptly and very patiently waited for the mechanical rabbit to come around again—and that's why we don't have cheetah races!

• •

Ice Age Souvenir Hunters

There is no doubt that Paleolithic man often migrated great distances as he followed the game herds. Evidence of his migrations is supported by numerous "souvenirs" he picked up along the way. Some of these objects may have been retained as records of his long and numerous journeys. They may even represent a record of contact of an unusual kind, just as vacationers currently collect objects as reminders that they have seen the mountains, the shore, or Rock City. Such mementos were frequently worn as decorative jewelry or amulets.

In the Lascaux Cave in France a rare variety of fossil mollusk shell was recovered from the floor by researchers. This particular fossil species is known today only from layered rocks on the Isle of Man. It had a slot in it and was probably carried on a sling and worn around the owner's neck. Was this a memento of his trip to a far-off land some 25,000 years ago?

Even more striking was that of a fossil trilobite found in another cave now known as the Cave of the Trilobite. It was also pierced and probably worn as an amulet. The ancient caveman who procured this souvenir really traveled, because the nearest source for this species of trilobite is in central Germany, nearly 1,300 miles to the east!

• •

The TV Roach

Scientists rank cockroaches among the most primitive of winged insects. Indeed their present-day appearance is very similar to that of their ancestors of 300 million years ago. To most humans they are repulsive creatures, probably because they abound in human habitations throughout most of the world. It is not easy to remain free of them, no matter how clean the human living quarters may be. They adapt quite easily to man's lifestyle and have now moved into one of man's more recent acquisitions in the home. This latter-day species is known as . . . the TV roach.

Supella supellectilium is a species of cockroach native to East Africa. With typical cockroach adaptability it somehow was transplanted to the United States and is now well established, particularly in New York City. The creature lives in TV sets, feeding on such things as paste, glue, and insulation. It never needs to come out of the set as do other household roaches, which feed on pantry provisions. Its TV home is cozy, giving it the darkness it prefers and tubes that provide heat. There seems to be no source of water, so in all probability this species metabolizes water as some desert insects do. This means the roach's own body is able to manufacture water molecules from the food it eats.

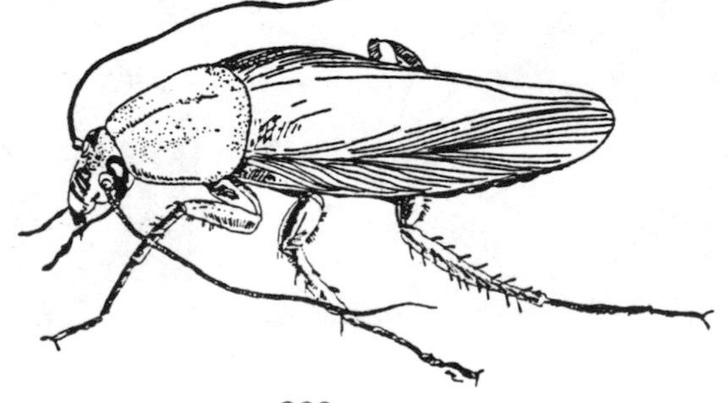

There is some merit to this species of roach, which shares man's habitat but not his habits. It does not eat man's food, sleep in his bed, or nestle on his body. Moreover it is not a disseminator of disease, and the actual damage it does to the television set is minimal. If this is not consolation enough, the TV owner may simply send the set's occupant to the Roach Motel!

• •

"On a Clear Day You Can See Forever"

Almost without exception, early motion pictures about the French Foreign Legion featured troops marching over endless rolling sand dunes. From this misleading view of the desert has evolved the impression that such topography is typical of all deserts. Actually sand dunes cover only a small portion of the deserts of the world. They make up a mere 3 percent of the great American deserts, and even in the vast Sahara only 30 percent is actually composed of dunes. The remainder, and by far the greater portion of these deserts, consists of rocky or mountainous terrain.

Anyone traveling across extensive desert areas immediately notices that the air is remarkably clear. The desert traveler can see, usually without

difficulty, all the way to the horizon. The major reason for the clear atmosphere is the low moisture content of the air.

Arid regions tend to have brighter skies than humid regions because water vapor absorbs some of the light passing through it. Therefore, with the scarcity of water in arid lands, it is not uncommon to see mountains more than forty miles away quite distinctly. Since such remote terrain appears quite clear and offers an unbroken view, distances are often misleading, and what appears to be only ten miles away may in reality be two or three times that distance from the viewer.

The atmosphere of the desert does, of course, contain moisture, but the proportion is relatively low compared with humid environments. Much of this moisture returns to the parched land in the form of dew.

Recently scientists have discovered that dew, the result of condensation of moisture from air, is an important source of desert water. It can often equal as much as ten inches of rainfall annually. Except for the driest of deserts, there is almost as much dew in arid lands as there is in the humid coastal regions! As a matter of fact, desert plants, as well as desert animals, supplement much of their meager ration of water with morning dew.

• •

"The Devil Dwells in Widecombe"

On the subject of tornadoes the midwestern United States has the dubious honor of being the world-recognized top achiever in all categories: number, severity, casualties, and damage. However, tornadoes have occurred at one time or another throughout the world.

Though less frequently, tornadoes have been recorded in the British Isles. Over many centuries they have occurred at the rate of about two every three years. The earliest authenticated tornado struck London on October 17, 1091, blowing down about 600 houses and churches. But in the entire violent saga of British tornadoes the most macabre tragedy of all occurred in a small English village.

The tornado struck on Sunday morning, October 21, 1638, at Widecombe-in-the-Moor in the Dartmoor region of Devonshire. The congregation had gathered in the church, and services had just begun when, without warning, everything went black and the building was hit by extremely violent winds and lightning. As the lightning struck, a ball of fire moved through the church and burst with a thunderous explosion.

Scores of people were killed and injured when the tornado struck the church. The roof and tower were wrecked, and rock and mortar showered down inside the building. People were snatched into the air, hurled against the pillars, and dismembered or smashed beyond recognition. The carnage was completed by the ball lightning.

Altogether the mayhem lasted only a matter of seconds, during which time over sixty people were killed. The benumbed survivors stood looking at the mass destruction with uncomprehending horror. The smell of the smoldering ruin, brimstone from the ball lightning, was to the people of that day the calling card of the devil himself. Many were thoroughly convinced that Judgment Day had finally arrived.

Centuries later a schoolchild in Devonshire was asked by his teacher, "What do you know of your ghostly enemy?" He replied, "If you please, ma'am, he lives in Widecombe."

• •

Prehistoric Ballast

Civilized man is often justifiably impressed by the high degree of awareness of their environment displayed by primitive people. An early explorer traveling on a South American river observed the crocodile's habit of swallowing stones. His Indian guides explained that they were retained in the crocodile's belly to improve stability and maneuverability underwater. The explorer was totally unconvinced, but scientists now know that the explanation offered by the Indians was absolutely correct. Modern discoveries suggest that this practice may have been established quite early among aquatic reptiles, especially for an extinct group of animals known as *plesiosaurs.*

These dinosaurlike reptiles were among the dominant marine reptilian species of the Mesozoic Era. They spent most of their lives in water, but the structure of their limbs clearly indicates that they did come out on land, probably to spawn. Their lifestyle appears to have been similar to that of the modern seal or walrus.

The limbs of a plesiosaur were usually about fifteen to twenty feet long. Although they were modified into flippers, they were reinforced by strong bone structures that would support the animal on land and give strength for paddling in water. They had two pairs of three-foot flippers, and it is believed they had huge lungs that gave them a natural buoyancy.

In Montana in the late 1980s two geologists discovered the intact rib cage of a plesiosaur; the living specimen would have been fifteen to twenty feet long. Aware that the air-breathing plesiosaur needed some way to counteract its natural tendency to float, they readily concluded that it could have counteracted this buoyancy by swallowing pebbles as does the modern crocodile.

Crocodiles would have the same buoyancy were it not for their intake of stones for ballast. This enables the crocodile to lie hidden underwater with only its eyes and nostrils protruding. In addition, the weight of the swallowed pebbles adds stability and helps it to drag large prey beneath the surface of the water.

The stone-swallowing habit of the plesiosaur was indicated clearly by the 197 stones found within the rib cage, ranging in size from a marble to a baseball. Since the rib cage was embedded in a fine-grained black shale that contained no other stones, there was no source for the pebbles other than the reptile itself. It is quite conclusive that this species of plesiosaur swallowed stones for ballast and was thus able to remain underwater to hide or catch prey or both.

When asked what may have killed this particular reptile some eighty million years ago, the scientists facetiously suggested that "doubtless it drowned after having eaten too many stones."

• •

"My Name Is Ozymandias"

The colossal temples at Abu Simbel in Upper Egypt might well be listed among the Great Wonders of the Ancient World. That architecture of such prodigious proportions could have been sculpted by man is, in itself, a wonder. The colossal statues and temples were designed and built to show the greatness of a single human being, the god-king Ramses II.

Ramses II was a spectacular king who briefly restored the glory of ancient Egypt. He ruled for over sixty years, 1304–1237 B.C., and lived past ninety years of age. During his reign the Nile Valley was studded with obelisks and colossal statues, mostly of himself, to commemorate his deeds. He was responsible for the Temple of Amon at Karnak in Thebes. It is the largest columned hall ever built and confirms, more than anything else, the passion to build in overpowering size and grandeur. The cost of building these monuments to glorify himself can never be estimated. How many slaves died laboring to depict his godliness is lost in the sands of time. He was indeed a tyrant obsessed with his own greatness.

Ramses II was not the first Egyptian king who had an obsession for self-aggrandizement. As would be expected of him, Ramses II appropriated many giant statues built by his predecessors, substituting his own name for theirs and incidentally usurping their deeds and victories.

The Greeks wrote of Ramses II, identifying him in translation as the great king Ozymandias and Thebes as "the world's greatest treasure house."

So impressed was the poet Shelley by the remains of the colossal statues that he wrote a sonnet that ranks among the immortal works of English

literature. An ironic commentary on the vanity and futility of a tyrant's power, it confirms that no matter how great a man may be he cannot withstand the ravages of time. Eventually he too will crumble to dust, which will be blended with the dust of the multitudes who died for his greatness.

Shelley's poem "Ozymandias" refers to Ramses II by the Greek version of his throne name. A portion of the sonnet is reproduced here.

> . . . "Two vast and trunkless legs of stone
> Stand in the desert. Near them, on the sand,
> Half sunk, a shattered visage lies, whose frown,
> And wrinkled lip, and sneer of cold command,
> Tell that its sculptor well those passions read . . ."
> And on the pedestal these words appear:
> "My name is Ozymandias, king of kings:
> Look on my works, ye Mighty, and despair!"
> Nothing beside remains. Round the decay
> Of that colossal wreck, boundless and bare
> The lone and level sands stretch far away.

Valuable Garbage

Before Maximilian became emperor of Mexico, he purchased two large diamonds. One was a forty-two-carat stone, later known as the Emperor

Maximilian, which he wore in a bag around his neck during his execution in 1867. It is now part of a private collection.

The other was a greenish-yellow fifty-carat gem that he bought in Brazil. He had it cut into a thirty-three-carat cushion-type stone and gave it to his wife-to-be, Princess Carlotta. It became known as the Carlotta.

The beautiful Carlotta diamond disappeared after his execution but turned up in 1901 when two men tried to smuggle it into the United States. The customs officials confiscated the gem and later auctioned it off for $120,000. In 1961, while in the possession of a New York jeweler, it met a rather unglamorous end.

The gem was secretly kept in a safe in the jeweler's home and zealously guarded. One evening when the jeweler's daughter was alone in the house she became alarmed because she thought she heard a burglar prowling around downstairs. The diamond was not at that time locked in the safe and was dangerously exposed. The frightened girl impetuously hid the diamond in a garbage pail! The burglar, as it turned out, was a product of her imagination. Nevertheless, no attempt was made to retrieve the stone until the next day. By that time sanitation men had already collected the garbage. The Carlotta, now part of the New York landfill, has not been seen since.

• •

The Deadly Wave—Tsunami

Named by the Japanese, *tsunami* refers to the large waves observed in harbors. They are also called *seismic sea waves* and, less accurately, *tidal waves*.

These destructive sea waves are generated by sudden movements of the sea floor resulting from earthquakes and volcanic eruptions. Compared with wind-driven waves, which seldom exceed sixty miles per hour, seismic waves may reach speeds of over 500 miles per hour!

Tsunamis travel across the ocean as concentric rings of troughs and crests, like ripples made when a pebble is dropped into a pond, but on a gigantic scale. The height and speed of the tsunami in open sea is determined by the depth of the water. In the deep ocean the length of the tsunami—the distance from crest to crest—may be dozens of miles, while the crest itself may be only two or three feet high. Because the wave is so low, ships at sea often pass over a tsunami without the crew's being aware of it.

When the deadly wave approaches land, the shallow water causes friction and slows up the front portion of the wave to as little as thirty miles per hour. The water behind it then surges forward with concentrated energy.

Thus the front of the wave builds up to tremendous heights, at times exceeding one hundred feet, and strikes the land with devastating force.

The highest recorded crest to strike land appeared off Valdez in southwestern Alaska on March 28, 1964, following the incredible Prince William earthquake, which reached a magnitude of 8.5 on the Richter scale. The wave generated by this quake crested at the unbelievable height of 220 feet.

When a tsunami moves into a funnel-shaped bay, its energy is concentrated and it grows to prodigious heights. Such deadly waves are not characterized by the spectacular breakers of storm waves, but instead they arrive as solid walls of water. They strike the land in a series of about eight waves. The first wave is often preceded by a sudden ebbing of the sea from the shore.

A particularly destructive wave hit the northeast coast of Japan on June 15, 1896. This tsunami was generated by a severe submarine earthquake with an epicenter about one hundred miles from the coast. At the head of a bay the waves rose to the awesome height of over one hundred feet. Devastating waves of this size swept an extensive area of Japan's northeast coast. It couldn't have happened at a more inappropriate time, because the beaches were crowded with people celebrating a Shinto festival. More than 27,000 people were killed, and well over 10,000 homes were swept away along a coastal area of over 150 miles.

Japanese fishermen returning home after a long day at sea were shocked and bewildered to find waters near the shore dotted with floating human corpses. At sea the tsunami had crested only one foot high, and the seamen were completely unaware of the violent waves that had passed under their boats. The same waves, unremarkable at sea, destroyed their families and homes when they struck the shore a short time later.

The most expert personal account of a tsunami was given by a scientist from Scripps Institution of Oceanography. He and his wife were staying at Kawela Bay on northern Oahu, Hawaii, when a tsunami struck on April 1, 1946. As the scientist left his house, he saw the water retreat suddenly for such a distance that the coral reefs were exposed. Numerous fish were left stranded on what had been, seconds before, the floor of the sea.

In a few minutes he could see the water beyond the exposed reef build higher and higher and higher until it suddenly shot forward with amazing velocity. He and his wife took refuge behind the house, which withstood the deluge, and were startled to see neighboring houses reduced to kindling.

As the wave subsided, the couple raced to higher ground before the next, much larger wave hit. There were then several more waves, but of smaller magnitude. A total of 173 people lost their lives. Many of the casualties were Hawaiians who went out to the exposed reefs to pick up stranded fish and were killed by the sudden onslaught of the second incoming wave.

The tsunami of 1946 was the last one to take Hawaii by surprise. It originated with a violent earthquake off the northern slope of the Aleutian Trench in the north Pacific Ocean. Five hours later, unheralded, the deadly waves struck Hawaii. Considering the amount of time between the earthquake in Alaska and the arrival of the tsunami in Hawaii, the island people could have been prepared if they had been forewarned.

As a result the tsunami of 1946 became the most important one in history, for it made the people aware that they need endure no more such destruction because of lack of warning. Action was taken that day by the U.S. Coast and Geodetic Survey, and work was begun on a Seismic Sea Wave Warning System.

Since then the islanders have been alerted to coming waves by a most efficient warning system. This was, however, not the end of deaths from Hawaiian tsunami. The fatalities that occurred on May 23, 1960, in Hawaii were due to a reckless curiosity on the part of the victims.

An earthquake off Chile had triggered the tsunami series, which raced across the Pacific at an average speed of 442 miles per hour. The first great wave was scheduled to reach Hawaii in fourteen hours and fifty-six minutes, certainly ample warning time to permit an effective evacuation. And so it did, with local people being moved to safety in the highlands. The tsunami was only one minute late!

Still sixty-one people died needlessly. A few of the evacuees tired of waiting in the hills and, thinking the danger was highly exaggerated, returned to their homes in the deserted villages. Most of the sixty-one people who died, however, had remained behind to "see the excitement." They lined the beaches and piers with cameras ready, and sure enough, they were treated to a firsthand view. Accounts related by the very few survivors labeled this one as an incredible act of foolhardiness.

• •

A New Type of Veterinary Medicine

There are times when the evolving nature of veterinary medicine can take a unique twist. This was emphasized in the treatment of Marvin, a 300-pound female Bengal tiger. Marvin, who was then an inmate at the National Zoological Park in Washington, D.C., was suffering from a three-and-a-half-inch laceration of her left leg.

Normally, treating the wound would appear to be a simple medical procedure, but the patient would not cooperate. On twelve separate occasions the zoo veterinarians had cleaned her wound with disinfectants and placed swathes of gauze bandages on the leg. This was a simple and effective treatment, except each time, in a few days, Marvin managed to gnaw off the bandages and reopen the wound.

Just before putting on the thirteenth bandage, the doctors had lunch at a nearby fast-food Mexican restaurant. Frustrated by the tiger's behavior, one of the veterinarians, acting on an idea, brought back with him a large plastic cup filled with hot sauce and chili peppers. After again bandaging the tiger's wound, the doctor smeared the hot sauce and peppers on the big cat's white bandage and, for good measure, added a touch of Tabasco sauce.

When placed back in her cage, the first thing she tried to do was to remove the bandage. One good taste, and her expression and reaction were almost the same as a human tasting something hot. Sticking her tongue in and out, she ran for the water tank and drank and drank and drank! This new approach to veterinary medicine was a complete success. Marvin left her bandage alone, and without further mishap the wound quickly healed.

• •

A Twist of Fate

To achieve a soft lunar landing that would permit man to walk on the moon, scientists and engineers realized that two space vehicles would have to be involved. The main vehicle would be taken into space and put into orbit around the moon. It would be carrying the smaller vehicle, which would later descend to the moon's surface.

Designing a small module to soft-land was much easier than landing an entire spacecraft. To return from the moon the crew of the lunar module would have to find the command ship and dock with it. Orbital rendezvous and docking were practical on paper but it had to be proved in space.

The first step was orbital rendezvous, and the first attempt was a failure. *Gemini 6* was successfully put into orbit around the earth, but the small *Agena* rocket scheduled to rendezvous never reached orbit. *Gemini 6* remained in space encircling the earth, and eleven days later *Gemini 7* was put into space orbit. It made a successful rendezvous with *Gemini 6*, proving the ability of spacecraft to meet and maneuver in space.

This maneuver completed, *Gemini 6* pulled away and returned to earth. *Gemini 7* remained in orbit for fourteen days, which was the amount of time the moon expedition was expected to take. Thus it was also proved that weightlessness was not a potential killer that would limit manned space programs but was well within the scope of human endurance.

The next major step was the actual docking of two craft in outer space. This was accomplished by *Gemini 10*. Project Gemini proved that man could live and work in space, certainly long enough to reach the moon and return.

Scientists and engineers now knew for certain that spacecraft could maneuver, rendezvous, and link up in space orbit. When *Gemini 12* splashed down in November 1966, it was clear that the race to the moon would be won by the United States.

During the phase of actual docking in space a near tragedy that could have changed space history occurred. The first docking in space was attempted by *Gemini 8* with its crew of Neil Armstrong and David Scott. No name in the space program is more memorable than that of Neil Armstrong, who was destined to be the first human to walk on the moon.

As the spacecraft moved up on the *Agena* target vehicle, pilot Armstrong managed to dock with the *Agena* rather smoothly. However, within minutes the two interlocked spacecraft began to tumble wildly. Armstrong, thinking the problem was with the *Agena*, quickly undocked the two spacecraft. But the problem was with the *Gemini*, and the spinning became worse; one of the control jets was jammed and not functioning.

Armstrong fought desperately to stabilize his spacecraft using thruster jets that are normally used for reentry into the earth's atmosphere. On the verge of blacking out, he was finally successful, and the spacecraft stabilized. *Gemini 8* returned immediately to earth with no further problems.

It seems ironic that the man who was destined to be the first man to walk on the moon almost became the first man to be lost in space!

• •

A Cushion to Fall Back On

Man is the only desert-dwelling creature that has not developed a natural means of surviving comfortably in the desert. There are, of course, exceptions to this rule, one of which is exhibited by the women of the African Bushmen tribe of the Kalahari Desert.

These women have developed a specific anatomical adaptation to desert living, a condition called *steatopygia*. It is well known that all human beings store food as fat, some more efficiently than others, and much time and money are spent on diets that are successful only for the duration of the program.

These layers of fat typically encircle the abdominal region and portions of the limbs. This fat becomes a great disadvantage in warm regions, because it is no easy task for a body engulfed in fat to get rid of excess heat. When a person moves or uses energy, heat is generated in his or her muscles. In desert regions, if there are fat layers in the limbs and abdominal regions, heat is retained, and it becomes very difficult for the person to stay cool. He or she could easily perish from heat prostration, as unwary desert travelers sometimes do.

The adaptation of the Bushmen women overcomes very effectively the fatty handicap suffered by other races and peoples, because their fat layers are concentrated in their buttocks. This part of their anatomy becomes huge and protruding and contrasts markedly with the rest of their lean, stringy bodies, which are devoid of fat. The exaggerated development of their buttocks reaches incredible proportions so that their girth at the hips often approximates their total height, which is less than five feet.

Doubtless these women appear strange and misshapen to outsiders, but they should be a source of envy to any sweat-drenched traveler who has visited the Bushmen's desert. No one could help noticing that these "deformed" women are relatively cool most of the time and rarely, if ever, sweat. Moreover, beauty being in the eye of the beholder, the Bushmen tribesmen consider these enormous buttocks a sign of great and lasting beauty and care little what the rest of her looks like!

• •

Carousel

Penguins, remarkable birds that live in temperatures of –60 degrees Fahrenheit, amid blizzards with wind velocities of up to ninety miles per hour, in the darkness of the long winter night within the Antarctic Circle, are among the most self-sacrificing and faithful creatures imaginable.

While other migratory birds move to warmer climates before winter, emperor penguins seek the most inhospitable region of the earth during its coldest, darkest, most hostile season of the year to mate, brood, and raise their young. On ice floes, usually near the same locale they visited the previous year, they assemble in great flocks after a trip of several thousand miles from any of the oceans of the Southern Hemisphere.

Upon arrival each male begins his courtship song. It is by this song, amid the cacophony of a thousand other mating melodies, that the female, most frequently his wife of the previous season, recognizes him. In about two months, during which time the birds have been fasting, an egg is laid. It is immediately transferred from the mother's brood pouch to the feet of the father, where it is covered by a protective flap of skin.

After a plaintive song of parting the female vanishes into the darkness to ice-free waters that are penguin feeding grounds. The male is left to care for the egg for at least two months. During that time he will partake of no food and will endure the harsh Antarctic winter with its furious, almost daily snowstorms and cold descending to –85 degrees Fahrenheit.

When a storm breaks, the fathers huddle and press together against one another, forming a sort of carousel that revolves slowly in a spiral. In this manner the rotating motion exposes each in turn to the raging wind while the others are protected. Each also individually migrates to the warm

carousel center. Only by this cooperative behavior are they able to survive the severest of storms. As soon as the storm subsides, the carousel breaks up but is quickly re-formed when a new storm approaches.

If, during the period of incubation, the male discovers the egg to be infertile, he will toss it away and head for the fishing grounds, but fathers with an egg to nurture remain faithful to their tasks. Encumbered as he is by the egg nestled on his feet, and emaciated from his months of starvation, the male would be defenseless against predators. Fortunately, and after generations of advance planning, the penguin has selected a season when no predators are about.

Most eggs do survive, and when the fledgling hatches, the father penguin feeds it a milklike secretion from his crop. The female returns, usually in June, with a gift of food for her husband; at this time she takes over the main activity of parenthood. During the next several months each parent will continue to spell the other in the rearing of their offspring. Only death can keep them from returning to their appointed duty. As the spring thaw breaks up the ice, young and old leave the colony for the open sea, there to regain their strength for the next season of child rearing.

CHAPTER NINE

So That's a Fossil!

Millions of years before man made his appearance on the earth, countless species of animals and plants flourished and dominated the land, the sea, and even the air. A record of such prehistoric life has been preserved

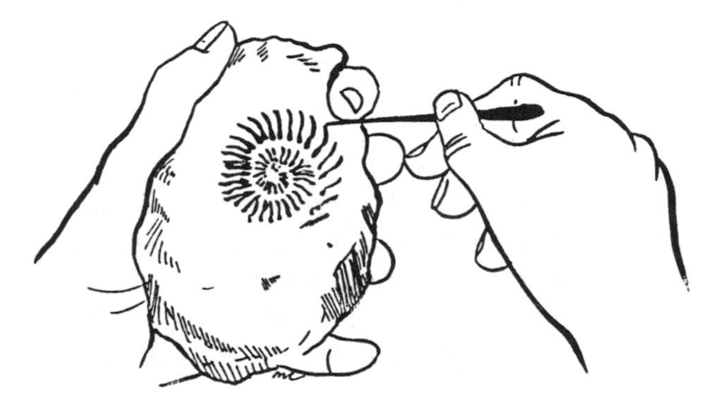

in the rocks that make up the crust of the earth. Scientists refer to the remains of these ancient inhabitants as *fossils*. But what things qualify as fossils?

The word *fossil* is derived from the Latin word meaning "to dig up." Early usage of the word in modern science included anything dug out of the earth, whether organic or inorganic. This usage undoubtedly led to much confusion, since "anything dug from the earth" could include a mineral or a rock or interred remains of practically anything within the category of fossil. As the sciences of geology and paleontology developed, the term *fossil* became restricted to include only those things that were remains or traces of organisms that lived during the geologic past.

Several years ago a show-and-tell geology student brought to class the mummified remains of a cat and rat that, he assured his fellow students, he had found the previous summer in the ruins of a Scottish castle. There was definite evidence to show that the animals had died in combat with each other. In the dry atmosphere of these ancient ruins the two animals had managed to avoid decomposition, and both were mummified in their entirety. Despite the remarkable state of preservation and the information that could be obtained from these specimens, they could not be classified as fossils, simply because they were not old enough.

In 1900 a hunter following a wounded deer in eastern Siberia happened upon the carcass of an elephant. The huge beast had been partially exposed by a recent cave-in of a river bank. The hunter's curiosity prevailing over fear and superstition, he approached the remains and fell to examining them. When he was satisfied with his investigation, he resumed his search for the wounded deer, but not before breaking off a tusk, which he later sold to a cossack. The buyer, upon his return to St. Petersburg, relayed the information that another frozen mammoth had been discovered.

The Imperial Academy of Science organized an expedition as soon as it was possible. The site of the discovery was at least 3,000 miles away, so it was more than a year later that the expedition arrived at the scene. By this time the exposed portions of the mammoth had undergone the ravages of two summer suns, and wolves had gnawed on its flesh. When the excavation finally began, the scientists were pleased to find that much of the carcass was still intact. Evidently the beast had fallen into a crevasse, breaking its hip. In a frenzied attempt to climb out it had pulled an avalanche of snow down on itself and was quickly frozen.

In its chest was a mass of clotted blood, and in the stomach was much undigested food. Between its clenched teeth there was food not yet swallowed, leaving no doubt as to the suddenness of its death. Frozen bacteria were later found in the mammoth's blood, suggesting that it must have had an infection at the time it was killed. The death of this unfortunate animal occurred some 45,000 years ago. The entire carcass of this mammoth, as

well as the bacteria in its blood, can be considered fossils. Why? Because they lived in the geologic past, well over 10,000 years ago.

Several hundred million years ago an animal crawled out on the soft mud near the town formerly known as Mauch Chunk, Pennsylvania. It may have been merely enjoying the afternoon sun, or perhaps it was on the never-ending quest for food. Whatever its reasons, it left behind perfect impressions of its feet in the soft mud, which was baked and hardened by the hot afternoon sun. The timeless processes of erosion and deposition covered and buried the footprints with sediments. Further geologic activity caused the mud to become sedimentary rock, and the footprints were preserved for aeons to come. Subsequent erosion in recent times stripped the overlying rocks and exposed several of the footprints to view. Although no single fragment of the animal is present, these footprints, as evidence of its having been there once ages ago, are therefore fossils.

In the summer of 1955, while prospecting for uranium in the vicinity of Price, Utah, a geologist and his party discovered the bones of a dinosaur. The remains were uncovered in sandstone from the Jurassic Period, indicating that this animal had lived over one hundred million years ago. It was identified as a swamp-dwelling, herbivorous dinosaur, the *Stegosaurus*. The bones were quite radioactive, a condition not uncommon in areas of uranium deposits. How it met its death was not evident, but the fact remains that the body was quickly covered by sediments and preserved. In time erosion removed the overburden and revealed this denizen of ancient days. The bones of this *Stegosaurus* are fossils since the animal lived in the extreme geologic past.

About fifteen miles east of the town of Wheeling, West Virginia, near streams along the old U.S. 40, is a well-known fossil plant quarry. The plants are nearly 300 million years old and are unquestionably fossils. Associated with the plants are petrified ripple marks showing evidence of a stream contemporary with the plants. Anyone who has glanced into a shallow streambed has noticed that the flowing water can produce ripples on the soft bottom muds. Most of the time these marks are readily destroyed by changing stream velocities. On occasion, however, the ripple marks are covered by subsequent deposition of sediments and are preserved. Ripple marks have been preserved in every geologic era; some of them are billions of years old. These ripple marks, which were preserved along with the plant life found in the quarry, are also 300 million years old. The plants are of course fossils, but the ripple marks, although they were formed in the same remote geologic era, could not be classed as fossils because they are not organic in origin. Mud cracks and raindrop imprints are also common evidence of events in the very distant past but, like the ripple marks, are not fossils.

Some years ago a geology professor welcomed a weatherbeaten wanderer with an Australian accent into his office. The guest was looking for a professional opinion of the contents of the box he was carrying. He opened six paper-wrapped packages, each containing a rock slab in which the remains of a perchlike fish were embedded. Although the professor had seen many fossil fish before, these caught his attention because the organic matter in each of the fish had been replaced with precious opal. The eyes were the most breathtaking—opalized gemstones the size of quarters, reflecting rainbow colors in the sunlight. The Australian wouldn't reveal the location of his find, nor has he yet responded to the professor's impassioned entreaties to take him along the next time he goes fishing. But somewhere in Australia . . .

Some forty million years ago, in what is now the Baltic states of Europe, there existed a species of tree that almost constantly exuded resin. Insect life must have been abundant, because the sap frequently descended on some unwary insect clinging to the tree or flying directly into the gooey material. The resin hardened around the encased insects and eventually developed into amber. Both the amber and its insect prisoner are considered fossils because both were organic and lived in the geologic past.

Despite the many ways in which fossils can be preserved, they must meet the requirements of being signs or remains of organic life from the geologic past. This category can include the merest fragments of bones and teeth or entire skeletons, shells, leaves or trees, an imprint, cast or mold that once encased ancient living matter.

Recent references to fossils have included "fossil fuels," those polluters of the biosphere that result from combustion of such materials as coal, oil, and natural gas. The term is completely valid, for every coal seam represents the remains of a carboniferous forest, and every pool of oil began as millions of microscopic animals from a prehistoric sea. Even the white cliffs on both sides of the Strait of Dover are composed of remains of Cretaceous (Lat. *creta*, "chalk") sea animals whose shells are visible under microscopic examination.

Only an infinitesimal percentage of life in the past has been preserved in the fossil record. Incomplete as it may be, it has provided indisputable evidence of the progression of life on our planet for nearly four billion years.

• •

A Modern Tale of Tigers

The history of man-eating tigers has filled many a volume. Some of the man-killers dined on several hundred victims before they were hunted down and destroyed. The main reasons for a tiger to turn man-eater have usually been old age or an injury that has handicapped the animal in its attempt

to capture its normal prey. Since its only alternative is to starve to death, the tiger turns on man, who, compared to its natural victims, is slow and puny—an easy catch. Once the tiger tastes human blood, it tends to shun all more elusive quarry.

There are other reasons for a tiger to become a man-eater. When a herd of cattle is attacked, the owner, in attempting to defend his cows, may also be attacked by the tiger. On such an occasion the giant cat would tend not to waste a meal and will dine on his human victim, thereby acquiring a preference for easy prey.

There have also been several reports of a villager's chance stumbling on a female tiger with cubs. The result is almost always the same. The tigress, out of fear that the human will harm her brood, instantly attacks. Such killings, referred to as "circumstantial," permit the tigress to discover that human beings constitute a rather effortless meal. She will then stalk and kill other humans for food, which she shares with her cubs. It is not unusual for her to teach her cubs the process of stalking and killing other humans for food. The cubs, having acquired a taste for human flesh, will grow up as confirmed man-eaters, and the females will repeat the process with each succeeding brood.

With the unrestrained expansion of the human population, the natural environment of the tiger has decreased, and the ranks of the giant cat have been depleted rapidly. The tiger was well on its way to becoming extinct in the wild. Alarmed conservationists have employed stern measures in some localities to preserve the species.

In the Indian Sundarbans, the marshy mangrove-forested delta of the Ganges extending across Bangladesh into India, conservation methods have been extraordinarily strict, and records have been kept. As a result of this effort the tiger population has nearly doubled since 1973. There are now over 260 of the giant cats living in the wilds of this preserve.

Many villagers have protested vehemently. Not without cause, it seems, for these people have always been the prime targets of man-eaters. With the rapid disappearance of the tiger they had experienced a reprieve, but as the tiger population began to rise the situation began to reverse itself. As a result the conservation program caused very bad feelings among the villagers. Their complaint remained the same: the tigers have shown they still have a taste for human flesh. In recent years they have killed as many as fifty villagers annually!

The officials of the Indian Sundarban Tiger Reserve have almost frantically cast about for methods to stop the carnage without having to kill the tigers. One of the managers believed that the salty marsh water that the tigers drank was the reason for their ferocity. Consequently many ditches were dug and filled with fresh water on the theory that if the tigers drink unsalted fresh water they will stop being so ferocious. This theory has now been abandoned and for very good reason: the attacks continued, if anything, with increased vigor.

In another attempt to protect the population the officials gave the people fiberglass protectors to shield their necks where tigers usually strike. This was another exercise in futility; according to the villagers the protectors were too hot to wear in the steamy climate. It is also very dubious that such a meager shield would stop a full-grown Bengal tiger.

Well, the authorities tried another method. They exploded firecrackers in an attempt to drive the big cats away from areas inhabited by people. This was quite obviously unsuccessful too, for while they were shooting off their firecrackers one tiger sneaked through the barrage and dragged a guard away.

It now appears that after many failures an effective method of discouraging the tiger from dining on humans is being developed. A series of very lifelike mannequins, dressed in villagers' clothing to give out a human scent, have been placed in various sites of tiger-human activity. There is nothing dangerous or menacing about them, except that upon biting into the human decoy the tiger is treated to a 230-volt electrical shock. This has been a very effective deterrent, and the annual killings have been cut to less than half of what they formerly were. Witnesses say that the tiger, when shocked, leaps back and roars frantically with pain. Then, running in circles with increasing speed, it breaks all records racing into the jungle away from the shocking "human." Thereafter it either flees people or definitely avoids the area where it was hurt.

While electrified dummies haven't ended tiger attacks, they certainly have reduced them. However, there's still no stopping a villager from defending his herd or stumbling onto a tigress with cubs. And so the attacks continue, a condition that may never cease until the distance between tiger and villager is increased or until one or the other departs from the wilds forever.

Here Comes the Bride

Nowadays when the wedding is over and the newly married couple hasten to start their honeymoon, in one final ritual the bride stops and tosses her bridal bouquet to her friends, who will scramble for it. This custom has its origins back in the days of ancient Rome.

In those days, when the bride entered her new home she was handed a torch. She promptly used it to light the hearth; then she tossed the torch to her friends, who would grapple for it. Sometimes she didn't even wait to extinguish the flaming torch. Whether it was lighted or not, the lucky woman who walked off with the torch was guaranteed exceptionally good luck for the remainder of the year. Understandably, the scrambling for the torch often became quite violent.

This custom survives today, with modifications. The violence is gone, and so is the torch. The bride instead tosses her bridal bouquet to her eager friends, and the winner is said to be the next to be married.

The Case of the Avian Thief

In the annals of crime one of the most unusual cases recorded was that of a bird that unwittingly succumbed to a life of thievery. This series of events occurred in 1938 in Chicago and left the reputation of the city temporarily tarnished.

A Chicago woman produced her own light-fingered mobster by patiently training her pet magpie to fetch shiny objects and bring them to her. Thus was launched the bird's career of crime. Residents of neighboring apartments reported missing jewelry almost daily to the police. The crimes were a real stumper. None of the victims could prove that their homes had been broken into, as doors were always locked and intact. Since their windows were several stories above the street, it appeared impossible for a thief to have entered from there. Still the thefts continued, and the police remained baffled.

Then one afternoon a potential victim stayed home from work with a touch of flu. She opened the window to let in fresh air and returned to bed. Rest was not to come easily, for as she lay abed gazing aimlessly, she was startled to see a magpie fly in through her open window. The woman watched, fascinated, as the bird flew around the room and settled on her dresser, where a diamond ring lay brilliantly reflecting the sunlight. Without hesitation the magpie seized the ring in its beak and flew out the open window.

Forgetting that she was sick, the woman ran to the window and watched the bird fly into the window of an apartment almost directly across the street. She quickly called the police and was eventually able to convince them that her fever had not caused her to hallucinate.

The police reluctantly "raided" the culprit's den. All were quite surprised at the rich store of stolen jewels the raid uncovered. The magpie's owner was promptly arrested and eventually drew a stiff prison sentence. As for the thieving bird, it made its "getaway" by flying out the open window shortly after the police entered and found the loot. They could only watch helplessly as the bird flew off into the sunset. To be sure, they had caught the real criminal, for the robberies ceased.

• •

Frog Dance

The celebrated croaking concert of the male frog during mating season is his only effort to attract a mate. The tactics he uses would be as effective for attracting prey, for as soon as a female, or a reasonable facsimile, swims by, he seizes her in a tight wrestler's grip and squeezes her until the spawn is expelled. Then he fertilizes the egg mass.

During the mating season it is not unusual for male frogs to yield to a form of mass mating hysteria. A male will jump on the back of another male, or a stone, or a piece of wood. They become so frantic that two males or more may jump on the same female. It is not unusual for a male to leap on the back of another male who is busily riding a female and ride him piggyback. Male victims usually struggle free of their attackers. Those who grab a stone or piece of wood drop the object as soon as they realize that it doesn't look, feel, or act like a frog.

In the midst of the mating hysteria it is not unusual for a male frog to grab a lump of mud. Apparently a lump of soft mud feels very much like a female, because he holds on very tightly. The mud becomes so modeled in his grasp that it assumes the modified shape of a frog. The male, deceived by the mud frog of his own creation, may hold on for days, waiting in vain for the captive lump to release spawn.

It is quite safe to assume that frogs have been around for well over 300 million years. During that time many species have developed very successful and unique methods of spawning.

In the land and water world of the amphibians the eggs of most species lack a protective membrane or shell to prevent the embryos from drying out. Hence the eggs must be laid in water, which introduces another crucial problem: how to keep the eggs safe from aquatic predators. The foam-nesting frogs of the African savannas have evolved an unusual solution. They keep their eggs safe from predators that might swim near by not laying them in water at all. For an amphibian this seems immensely foolhardy, but the foam-nesting frogs have worked it out rather well.

During mating season both males and females gather in a pond. They usually emerge in groups of up to forty, climb a nearby tree, and crawl out on a branch overhanging the pond. Once they are out on the limb the females secrete a mucus on it. Immediately all the frogs collectively begin to kick madly with their hind legs, whipping the secretion into a froth. What a sight to behold—forty or so frogs all in a frenzy, performing a ritual dance as they kick the mucus into a white, foamy meringue! Once the froth is well developed, the frogs spawn in it and lay their eggs.

Immediately after the eggs are laid, the frogs depart, their mission in life now completed. The froth quickly hardens on the outside, protecting the embryos from dehydration and very likely also from land-dwelling predators. In about four days the well-developed tadpoles begin to emerge from their eggs. By wriggling sufficiently they manage to make a hole in the outside shell of the froth. They emerge from it and drop several feet directly into the pond below. Their life as free-moving vertebrates has now begun.

• •

A Classic Split Personality

There are many strange cases of multiple personality, but none more strange than that of William Brodie. He was a highly respected man, son of a wealthy cabinetmaker, a member of the city council, and deacon of the masons' guild. He led a most admirable life—that is, until August of 1768, when he began to lead two very different lives. By night Brodie was the leader of a vicious gang of thieves and murderers. His nighttime crimes were at times unspeakably sadistic. The man was absolutely without compassion and reveled in the misery of his victims.

By day it was a completely different story. Deacon Brodie, leading citizen, was kind and generous to most of the people he encountered. It is even doubtful if he was fully aware of what kind of man he became at night; in fact his most probable link with his late-night behavior was an occasional dreamlike, fleeting remembrance. Absolutely nobody knew of Brodie's secret night life—not even his two mistresses—and he maintained this dual existence for over eighteen years.

He was finally caught when he and his gang tried to break into the Scottish customs and excise office. The public was absolutely unbelieving that such a highly respected man was the almost legendary leader of a gang of cutthroats.

People came from miles around to attend his trial, mainly to see for themselves what Deacon Brodie looked like. They nearly always left in disbelief that such a kind-looking gentleman could be the leader of thieves and murderers. Evidently the jury had another opinion, because he was quickly found guilty and subsequently hanged.

In 1884 Robert Louis Stevenson and William E. Henley wrote a play based on Brodie's career called *Deacon Brodie or The Double Life*. Stevenson went a step further and used Brodie as the basis for a new book he wrote in 1886. He entitled it *The Strange Case of Dr. Jekyll and Mr. Hyde.*

• •

Destination by Instinct

Some bird migrations are true wonders of nature; that of the young bronze cuckoo of the southwestern Pacific is a baffling accomplishment. Like many species of cuckoos, it is a parasitic breeder. The mother, in New Zealand at nesting time, lays her eggs in another bird's nest and promptly forgets about them. Her most responsible act of parenting is to select a nonmigratory bird, such as the flycatcher, to be the foster parent to her young. As additional insurance of their well-being she often dumps some of the foster parent's eggs overboard to make room for her brood.

Then the bronze cuckoo spreads her wings and heads for the Solomon

Islands. Meanwhile the foster parents don't seem to observe the difference in the cuckoo eggs. They incubate the foreign eggs, feed the fledglings, and eventually teach them how to fly.

When the young cuckoo is fully mature, it leaves the nest of its birth and sets out on a journey that carries it 1,200 miles across the sea to Australia. There it rests briefly and then continues its flight. It will then cross another 1,000 miles of open sea to the Solomon Islands. Here it will join flocks of cuckoos that had left New Zealand many months before it had even hatched from its egg. Somewhere among the adult birds in the flock may be the cuckoo's natural mother.

Most amazingly the young cuckoo is able to make a completely unguided flight across over 2,000 miles of open sea. Both the flight path and the destination could have been reckoned only by instinct. After that, finding its own flock must have been child's play.

Experts in bird behavior do not regard the penguin as a creature of high intellect, and this opinion is well founded. If it were taken out of its natural environment, the penguin would starve to death unless force-fed or hand-fed for a period of time. It would be unable to locate food in a fish market! Being preprogrammed, the penguin does not normally face situations in which logical reasoning is necessary.

As a powerful redeeming feature the penguin has an unshakable homing instinct. Obviously, then, penguin behavior is based on instinct rather than a rudimentary capacity to reason. This would account for its ability to navigate unerringly over vast reaches of open ocean, devoid of any landmarks and quite often fogbound.

As an experiment, early in 1959 scientists captured five adult Adélie penguins near Wilkes Station in the Antarctic. The captives were banded with sensitive tracking devices and flown to McMurdo Sound, over 1,200 miles away. Here they were released and tracked carefully to see if they could make it back to their home ground.

The following spring the scientists returned to the Wilkes Station rookery as the penguin population assembled for mating. The former captives waddled ashore and asserted claims to the very nesting sites from which they had been forcibly removed—ten months previously.

• •

Man's Humanity Toward Man

More than half a million years ago a human quenched his thirst at a nearby stream. Whether he was careless, inattentive, or momentarily off guard, his leg was suddenly seized in a viselike grip by a gigantic crocodile. The man was in the process of being dragged into the river when his

screams attracted his companions. They ran to his rescue and assaulted the reptile with rocks and clubs until it was forced to release its victim and swim away. The unfortunate human was carried to camp, where he lay for some weeks while his leg slowly healed. His injuries disabled him for life.

The interpretation above is based on a recent find of a *Homo erectus* leg bone in Java. It showed evidence of a very disabling crocodile bite that later healed. The nature of the wound clearly showed that the man had been crippled permanently, yet he lived for years after the incident. To be unable to fend for oneself in the hostile world of the Stone Age the man must have been cared for and protected by his companions.

Another recent find in Kenya shows signs of human compassion within a family group. This again involved a *Homo erectus* and occurred in the same geologic age as the incident described above. The find was a diseased partial skeleton of a female.

Scientists studied the fossil bones intensely and determined that the cause of death was a bone disease resulting from an excessive intake of vitamin A. At the time that she lived, about 500,000 years ago, the easiest way to get a whopping dose of vitamin A was to eat the liver of a large carnivore. All animals store this vitamin in their liver, and many species have the metabolic ability to partially detoxify it. Humans do not have this capacity. Since carnivores always eat the livers of their kill, their intake of vitamin A is enormous. Therefore, in the course of its life a carnivore's liver becomes more and more toxic.

What is most surprising in this case is that the woman was able to obtain an entire carnivore liver, or at least a large portion of it. Considering that the liver is not likely to be left to scavengers, this liver was most certainly the prize from a fatal clash between a family of *Homo erectus* and a large carnivore, such as a saber-toothed tiger. The latter lost!

The woman was probably present during the fight, but how she was permitted to keep and eat a large share of the liver will certainly never be known. As she greedily enjoyed the liver, she was unknowingly condemning herself to a slow and painful death. The effect of eating such a vitamin-rich liver was dramatic; it adversely affected her entire skeletal system. Examination of the remains showed evidence that this Paleolithic woman must have lived for several weeks after she consumed the toxic liver.

To have lived so long in a state of physical disability and pain, she would most certainly have become quite helpless. She must have been cared for by her companions, presumably members of her family or tribe. There is no doubt that early man was a ruthless predator, but he was also gregarious and emotionally attached to his family.

A recurrent thread of information acquired from the study of Neanderthal remains is that their social bonds were exceedingly strong. Consider the plight of the "Old Man of La Chapelle-aux-Saints," a predicament quite similar to that of the Shanidar caveman cited earlier in this book. The "old man" had died over 50,000 years ago and was definitely given a ceremonious burial. In life he had suffered a broken rib, severe hip arthritis, and diseased vertebrae, and to make matters worse he had a gum disease that caused almost all of his teeth to fall out. He was certainly unable to hunt; he couldn't chew very well, and it is doubtful that he could even walk without some kind of support. Yet he survived to the ripe old age of forty, a real senior citizen among the Neanderthals. Clearly his companions had to provide for and take care of him or he would have died at an early age. His usefulness was probably restricted to menial, sedentary jobs, such as tending the fire.

Since most Neanderthals died in their early twenties, it seems ironic that, because he was relieved of hunting responsibilities, he was exposed to few of the hazards of the Stone Age and therefore outlived most of the people who took care of him.

The incident above and a host of others reveal altruism and a social consciousness that one would scarcely expect from the brutish, savage Neanderthal. As one scientist pointed out, "In the light of twentieth-century human behavior, we should be careful whom we call brutish."

• •

Reincarnated Rembrandts

Captive chimpanzees have been taught a variety of skills, and some have learned a few tricks on their own. At least one chimp has taken to the fine arts.

In 1964 there was an exhibition of the art of a Monsieur Pierre Brassau, which drew much praise from the critics. Of course they were unaware that Monsieur Brassau was a chimp named Pierre, who created his art while seated before his easel eating a banana and wielding a brush and palette in his cage in a Swedish zoo.

Actually the paintings were somewhat messy and elementary. But to trendsetters of the sixties they were attention-getting and showed the genuine originality demanded during this era of pop art. Pierre did, however, have competition.

San Francisco, the birthplace of many different forms of psychedelic art, was also the home of a natural artist known to his associates as "Willie." It is not particularly surprising, therefore, that Willie, who executed some most unusual paintings, was a large earthworm.

Willie's owner would dip him in paint, drop him on a canvas, and let him wiggle his way across. Multicolored paint provided a most colorful mess of meaningless patterns. This went on for nearly two years, 1963-65, during which time Willie produced nearly 200 paintings, some selling for up to $100. There is no doubt that beauty is in the eyes and mind of the beholder!

But is it art? If art is "the perfection of nature," "the embodiment of

great ideas," or "the immortal movement of its time," the works of Pierre and Willie would scarcely qualify. If art can flourish only "when there is a sense of adventure . . . of complete freedom to experiment," the directors of Pierre and Willie are the artists. But if art is the "most intense mode of individualism that the world has known," quite possibly the chimpanzee and the earthworm are among the greatest.

• •

Cullinan Diamond

The largest diamond ever discovered has not had a history steeped in intrigue, treachery, and death. Instead the events that surround this fabulous find are infused with mirth and good-natured levity. This is the story of the great . . . Cullinan Diamond.

It was in December 1902, just seven months after the Boer War had at last brought the Boer republics under British sovereignty, that a one-time bricklayer named Thomas Cullinan discovered diamonds on a small ground rise near Pretoria, the administrative capital of the Transvaal. This little hill was found to be the cap of a new diamond pipe. It became known as the Premier Mine, and within a few years it was yielding more diamonds than all the mines controlled by the fabulous DeBeers enterprise.

That such a competitor was permitted to exist represented a definite shortsightedness on the part of the DeBeers chairman Francis Oats. Convinced that Thomas Cullinan had salted his hill with diamonds from elsewhere, Oats concluded that he could safely sit back and await the inevitable embarrassment of those who had invested in the mine. At the very beginning he might have been able to acquire the Premier for a song. He soon came to regret his decision not to buy, for the mine continued to yield more and more diamonds. To add insult to injury the directors of the Premier sold their diamonds on their own, refusing to cooperate with the diamond syndicate that had been set up by DeBeers to control the world's diamond trade. But the greatest blow to DeBeers was the discovery in the Premier Mine of the largest diamond the world had ever seen.

It was late one afternoon in January 1905 when the mine superintendent, Frederick Wells, was conducting an inspection tour. He could scarcely believe his eyes when, protruding from one of the side walls and reflecting in the setting South African sun, he caught sight of the largest raw diamond he had ever seen. He scrambled down into the thirty-foot pit and, using his penknife, dislodged the gemstone. It filled his hand!

The stone had three natural crystal faces and one large cleavage. The nature and shape of the cleavage have led many scientists to believe the stone was probably part of a large lump of diamond that had broken loose during earth movements. Indeed some authorities recognize the remarkable possibility that the total raw diamond lump may be twice the size of the fragment found by Wells. The stone, in the rough form in which it was discovered, measured four-and-a-half by two-and-a-half inches and weighed 1.4 pounds; this is an incredible 3,106 carats. It was named the Cullinan Diamond after the chairman of the Premier Mine. The discoverer of the gemstone, Frederick Wells, was given a bonus of $10,000 dollars, a considerable sum in those days.

From the moment of its discovery the Cullinan was treated as a stone beyond compare. The Premier Mine sold the diamond to the government of the Transvaal for £150,000. They in turn presented the stone to King Edward VII of Great Britain on his sixty-sixth birthday in 1907 as a token of colonial allegiance to Great Britain.

The mirth associated with the stone arises when one learns of the method by which it was sent to the new owner. The government of the Transvaal, being concerned about the possible theft of so precious a cargo, employed very complicated security measures in shipping a box, supposedly carrying the stone, to London. However, it was all a subterfuge. While making a grand display of transporting the diamond by means of a highly protected operation, it was in fact sent by parcel post bearing a three-shilling stamp.

Then comes part two of the comedy. After much consultation King Edward appointed the Asscher Brothers firm in Amsterdam to cut the stone. Another ruse was employed in transporting the stone from London to Amsterdam. Segments of the Royal Navy transported an empty box across the North Sea. Even the captain of the actual transporting vessel did not know that the closely guarded box in his cabin was empty. Meanwhile, Abraham Asscher traveled home with the Cullinan Diamond in his pocket.

Abraham's brother, Joseph, undertook the task of cutting the diamond. Awed by the possibility that the great stone could be ruined, Joseph studied it for several months before attempting the first cut. It was on February 10, 1908, that he finally placed a steel blade in a newly made incision and rapped it firmly with his mallet. The diamond did not cleave; instead the steel blade broke. Joseph made three more attempts before the diamond finally split. To the amazement and relief of all concerned it divided perfectly into two parts.

Associated with the cleaving of the Cullinan is a persistent melodramatic legend. It is said that when the stone cleaved successfully Asscher swooned into the arms of his doctor.

Asscher subsequently fashioned the Cullinan into nine large and ninety-six smaller stones, as well as fragments totaling 9.5 carats. Presently two of the major gems are the largest cut diamonds in the world. The Great Star of Africa is the largest, weighing 530.20 carats. It is mounted on the British Imperial Sceptre and is a permanent display in the Tower of London.

The second gem is generally spoken of as "The Cullinan." It is set in the Imperial State Crown. Like the Great Star, Cullinan II is one of the British Crown jewels. All nine of the largest stones fashioned from the raw diamond have remained in the possession of the British Crown.

It seems refreshing that the largest diamond of all time does not have a history of tragedy or intrigue but one of humor and majestic poise. Such a noble and respected gem deserves to be where it is today!

• •

Suddenly the Sky Darkened

In 1980 the fossil remains of an eight-million-year-old bird were unearthed in Argentina. Its most unique feature was its wingspan of a least twenty feet, making it the largest known bird to fly. The great teratorn, a vulturelike predator, has been named *Argentavis magnificens*. Measurements taken of its wing bones clearly show they are the right size for its body proportions to have achieved true flight. The bird was incredibly large; standing on the ground, it could have looked a six-foot man straight in the eye!

Prior to the discovery of this new species, the largest known bird capable of flight was the vulture *Teratornis*. Several skeletons have been found in the La Brea Tar Pits of Los Angeles. It had a wingspan of twelve feet and stood thirty inches high. Next to the new Argentina specimen, however, it would appear to be a diminutive juvenile!

• •

Nemesis!

Binary star systems are not unusual in our universe, and some scientists accept the possibility that our sun is a member of such a system; that is that our sun has an as yet unseen companion star. It is conjectured that this hypothetical star has an elliptical orbit that brings it close to the solar system approximately every twenty-six million years.

Astronomers believe that a cloud of virtually billions of comets circles the sun in an orbit far beyond that of Pluto's. It is speculated that when the approaching star gets close to the solar system enough gravitational pull is exerted on the comets to disrupt their orbits. This causes a few of them to careen toward the sun, creating the possibility of one or more of them impacting with the earth.

Mass extinctions of plant and animal life are believed to result from such collisions. Supporting this belief is the discovery by scientists that world-

wide major biological extinctions do seem to occur approximately every twenty-six million years. There is growing evidence to indicate that dinosaur extinction may have resulted from the collision of the earth with a gigantic body from outer space.

Whether the extinction theory is true or not, it is small wonder that proponents of this hypothesis refer to this star as Nemesis!

• •

Life-and-Death Deception

Ways in which animals deceive an enemy or intended prey are numerous and usually quite successful. Some appear to border on the realm of science fiction.

A rather rare phenomenon was observed by several scientists in the Indian Ocean, where they were observing a large school of tiny fish, each no bigger than a finger. Suddenly a large predatory barracuda approached. A quiver was seen to run through the school, and instantly the tiny fish closed ranks and, as if they were one, assembled a formation that strongly resembled an eighteen-foot shark. The scientists observed that four times in succession this sea monster composed of thousands of tiny units leapt high in the air as if it were a single organism, then splashed back into the water dolphin-fashion. The barracuda, seeing its prey so abruptly transformed into a huge shark, paused for a moment "as if it had to wipe its glasses." Then, without hesitation, it fled from this life-threatening monster.

Animal life-and-death deception can also work in favor of the predator. The African mongoose is a deadly enemy of the male francolin partridge that dwells in the bush with his harem of hens. The reason the mongoose is such a treacherous enemy is that it can imitate the male francolin's cry perfectly. This talent enables the mongoose to fool the male partridge, for when it discovers a flock of these birds in the vicinity it crows like a cock partridge. The real harem boss thinks that a strange male francolin is in the bush and wants to fight him for his harem, so he rushes toward the sound in reckless attack posture. The mongoose needs merely to wait as his meal rushes right into his open mouth. Dinner is served!

On the other hand, the mongoose also falls victim to a life-and-death deception. Ordinarily, when encountering a snake, the mongoose wins out, but not always. There are some vipers whose tails confusingly resemble their heads, so when the mongoose attacks, the snake raises its tail threateningly as if it were confronting the mongoose. The mongoose is deceived effectively and bites the imitation head. Then, within a fraction of a second, it is itself bitten by the poisonous snake's real head, complete

with fangs and venom. The mongoose that planned to dine on the snake instead becomes the dinner.

But here's a tale of the champion reptilian deceiver. In July 1951 a graduate student in geology was part of an early man archaeological dig in central Nebraska. His job was to identify the fossil bones as they were excavated. Being young and full of vigor, he spent the weekends prospecting for older fossils, which he stored at camp, awaiting the time when they could be shipped to the University of Nebraska, Lincoln. The unusual event happened late one Sunday as the geologist, his stamina dwindling after spending an entire day in the field by himself, was returning to camp.

The geologist suddenly stopped in his tracks, for lying on the trail in front of him was a snake that appeared to be of enormous length. The snake reared up the front part of its body, which promptly inflated into a large hood, and proceeded to hiss violently in a frightening manner. It scared the dickens out of the budding scientist, who exclaimed, "Ye gads, a cobra!" He then proceeded to clobber the snake with the shovel he was

carrying. The snake was quickly sent off to reptile heaven while the man ran to camp with his story of having just killed a giant cobra.

All of the expedition members accompanied him back to the site of his encounter with the killer snake, and there lay a two-foot hognose snake. The ribbing he took continued for the remainder of the expedition—almost two months!

The young man was not the first to be frightened by the antics of this type of serpent. The Nebraska hognose snake in reality is a harmless, nonpoisonous snake that puts on quite an act on certain occasions. When it encounters danger, it inflates the front half of its body to about twice its normal size, spreading and flattening until it resembles a cobra, as could be verified by the geologist. Once the front end is "fanned out," the snake strikes vigorously amid loud hissing. These actions are certainly enough to discourage its being bothered further by most people, provided they are not carrying a shovel.

If mimicking the cobra does not work, the snake seems to go into convulsions, flips over, and goes perfectly limp as if dead. If the intruder attempts to make sure the snake is dead by turning it right side up, the hognose will immediately flip over again on its back, emphasizing the fact that it is indeed dead.

On occasion the "dead" snake will raise its head to see if the danger has passed. If not, the head quickly drops back down again into the dead position. When the danger has really passed, the snake will flip over on its belly and slither away.

The hognose snake should definitely be nominated for an Oscar as best reptilian actor in a short subject.

• •

How Custer Almost Missed His Massacre

About a decade before the Battle of Little Bighorn, General George Armstrong Custer, while hunting bison on the Great Plains, was dismayed to see a large bull turn and charge his horse. The general must have panicked, because he accidentally discharged his gun, shooting his horse through the head and killing the unfortunate mount instantly. Custer, who was actually a colonel at the time, was rather unceremoniously dumped to the ground. The bison must have thought this to be inappropriate behavior for a prospective victim, because he dropped his attack, turned, and walked away.

• •

The Hawk—Nature's Air Force

The graceful hawk is one of the most beneficial of living predators because of its effective performance in maintaining the balance of nature. It preys on fast-breeding creatures such as mice, rats, squirrels, rabbits, and insects. If the population growths of these animals were to go uncontrolled,

it is not science fiction, but grim reality, that eventually they could inherit the earth. So efficient is the hawk that systematic studies have shown that a single hawk killed over fifty meadow mice on one-quarter of an acre in thirty days.

The hawk also preys on fast-breeding small birds such as sparrows, chickadees, and numerous other species. Without the hawk and related predators to control their numbers these fast breeders would eventually become a dangerous nuisance and perhaps even a national hazard.

From time to time problems have arisen with unchecked population growths of such animals as various species of rodents. When investigated, it is usually shown that the local hawk population has been decimated by overzealous hunters. Thus, with little to keep the rodents in check, a population explosion occurs. For man to inadvertently disrupt the balance of nature is, of course, nothing new.

The ease with which the hawk takes its prey illustrates perfect coordination of eye, wing, and talons. Hawks have been observed to swoop over a pond and pick off a moorhen without even making a splash. They have been seen plucking lizards from tree trunks without even a pause in flight. They are able to capture small birds in flight so swiftly that no act of seizure could be detected.

Recently a scientist observed the split-second timing of this predator. In this case the hawk was in pursuit of a quail. The hawk was definitely gaining when the quail suddenly dropped like a rock toward a bush. The hawk then hurled itself through the air, flung its body beneath the prey upside down, and received the falling quail in its talons. Righting itself, the hawk flew away with its prize.

Tempest in a Teapot

Many years of documentation have confirmed that the most hail-pummeled place on the planet is the Kericho Hills region of Kenya in East Africa. This area undergoes hailstorms on the unrivaled average of 132 days per year. This is also the area of Kenya's tea plantations, which yield quite plentiful and profitable crops, placing Kenya among the top ten tea-producing nations in the world. The severe localized hailstorms often cause considerable damage to the tea crop and much concern among plantation owners. In 1978 a team of scientists conducted an investigation and reported that the link between hail and tea is not coincidental. In fact their research clearly indicated that there is a cause-and-effect relationship between the hail and the tea.

Organic litter from tea plants is strewn on the ground between hundreds of rows of the shrubs. Much of the litter is composed of dust-type particles of just the right size to serve as seeds or nuclei around which hailstones can grow. The activity of hundreds of tea pickers employed by the estates readily churns the tiny particles up into the atmosphere. The atmosphere in turn becomes choked with organic dust particles, which can serve as nuclei for hailstones and provoke the storms.

The conclusion of the researchers was that these recurrent hailstorms are induced organically. As one of the scientists observed, the tea produces the seeds for its own destruction!

• •

Moray!

The moray eel, according to early divers' stories, is one of the ogres of the deep. As fish stories go, the moray's reputation is far worse than it deserves.

At present 120 known species of moray eels are living in warm seas, especially around coral reefs. To fish for the moray one would have to seek out tropical or subtropical waters no deeper than 150 feet, because eels are rarely seen in the open sea. They range in size from a mere six inches to well over twelve feet in length. Undoubtedly the long, snakelike body has contributed to its negative reputation, and some recent motion pictures have capitalized on just that fact.

The Romans must have regarded them as an esteemed delicacy. There is a record of a single banquet given by Caesar in which at least 6,000 moray eels were eaten as the main course. They have been eaten in Mediterranean countries ever since.

As part of their bad reputation it is believed by many that the morays are venomous. But so far there is no evidence to support such a conclusion.

They also have the reputation of being aggressive, of attacking bathers and divers and people searching the reefs for lobsters, abalones, and other shellfish. They are believed to be capable of taking a tenacious grip on a man's arm and holding him underwater until he drowns. Contrary to all this, present-day evidence compiled by skin-diving scientists clearly shows morays will try to avoid humans, probably more anxiously than any person tries to avoid them.

Not surprisingly, when cornered or speared, a moray eel will lunge and bite in a tremendous effort to escape. There are authentic records of severe wounds resulting from encounters with this fish. In attacking, the moray's behavior seems to resemble that of a venomous snake, for it rears its head and the front part of the body and strikes down. These movements undoubtedly reminded early divers of a striking snake.

However, make no bones about it, there have been cases of people being attacked by morays that were not provoked. But scientists strongly believe this occurs only during breeding season. Sexual excitement apparently drives them into a frenzy. When the *Kon Tiki* raft was wrecked on an atoll in the Pacific, its crew was chased from the lagoon by morays. The scientific members of the Royal Indian Marine Survey ship *Investigator* had a similar experience on the Betrapar Atoll in the Indian Ocean in 1902. On that occasion it was noted that the eels were breeding.

There is a growing suspicion that many sea serpent stories are founded on the sightings of commonplace objects seen at unusual angles at various times and places. One could be the moray eel. When disturbed, the moray will sometimes swim at the surface with the forepart of the body and head held high above the water. A twelve-foot moray with its bizarre coloring, swimming in this manner with the front end of the dorsal fin looking like a mane, would approximate the conventional picture of a sea serpent!

Morays have a trick of throwing their bodies into a knot and letting this knot travel forward to or backward from the head. This is a useful trait. A scientist recently observed an octopus grab a moray about the head. By throwing itself into the movable knot it slipped its body back through the loop to force the octopus's tentacles off its head. When a moray eel is hooked on a line, it will try to free itself in the same manner. And somehow, in so doing, the eel actually climbs the line, tail first.

This maneuver can have terrifying results, as demonstrated by a recorded incident off Palm Beach, Florida. Here three fishermen in a relatively small boat were enjoying an afternoon of fishing until one of them hooked a twelve-foot moray eel. In trying to escape via the knot route, the fish climbed the line. The three men were paralyzed with fear as a gigantic moray eel came writhing aboard their boat tail first! Without hesitation all three fishermen jumped overboard!

• •

Ice Age Lion

The most famous of the Ice Age predators to inhabit California was the large saber-toothed cat *Smilodon*. A more formidable and larger specimen of the carnivorous cats, however, was the gigantic American lion *Felis atrox*. The animal was proportioned very much like the modern African lion and probably looked quite similar, with one major exception—it was more than one-third larger and heavier!

The famous saber-toothed cat was about the size of a modern African lion and would therefore have been dwarfed if it stood next to the Ice Age lion. Carnivore remains found in the La Brea Tar Pits in the Los Angeles area number in the thousands and are represented mainly by the wolf and the saber-toothed cat. Obviously these carnivores ignored the death trap of the ages and seldom hesitated to rush into the tar pits to secure some helplessly trapped horse or deer. The record of almost 100 percent fatality to all that trespassed there never seems to have pierced their thick skulls.

Approximately one hundred skeletal remains of the American lion have been recovered from the La Brea Tar Pits. Compared to the thousands of saber-toothed cats recovered from the pits, this is a remarkably small number. It can be assumed that they had a high degree of intelligence and learned to avoid the death trap beckoning to them from the tar pits. Despite its higher-level thinking, the Ice Age American lion did not outlive the saber-toothed cats. When the great Pleistocene extinctions took place, the American lion was right in there!

• •

"Ring Around the Rosy"

In preschools throughout the country one nursery song is among the first in toddlers' repertoire of circle games. A group of small children will, with very little prompting, form a ring, join hands, and dance in a circle, singing merrily at the top of their lungs, "Ring around the rosy."

This song comes from a time when death was the byword in Europe. It was the time of the Black Death.

In the mid-fourteenth century a devastating plague struck Europe. It received its name from the black spots, or buboes, that appeared on the skins of the victims as the result of blood hemorrhages under the skin. The spots were always surrounded by large, distinct reddish rashes. The Black Death was bubonic plague, caused by the bacterium *Pasteurella pestis* and carried by fleas and rodents. The plague entered Europe by means of rats on merchant ships, arriving in southern Italy in 1347. It quickly spread, via the trade routes, to Spain and France. It reached England in 1348, Germany in 1349, and Russia in 1350.

When the disease struck, Europe was in a most helpless state to combat it. The people had no natural immunity to the disease, and standards of public health and personal hygiene were, to say the least, quite low. Medical science and the clergy couldn't cope and, whenever possible, retreated from contact with infected people. In the absence of any known cause, people arrived at the inescapable conclusion that it was a pestilence visited by an angry deity on a very sinful world.

No one is sure just what the mortality rate was, because very poor records were kept during the climax of the plague. Since the records are unreliable, scientists can only estimate how many people died. It is agreed by authorities that anywhere between one-quarter and one-third of Europe's population died between the years 1347-1350 as a result of the Black Death. Incredibly this means that over twenty-five million people succumbed to the disease.

It was in England that the nursery rhyme originated. But certainly not for childish play. During the high point of the plague people became so accustomed to sickly death that mass hysteria was not uncommon. The populace gathered frequently in the streets and danced hysterically in anticipation of "tomorrow we die." Dancers leapt, screamed, and sang phrases to the overshadowing Black Death. People of all ages often danced and sang until complete exhaustion took hold and the revelers dropped to the ground. There is little doubt that the frantic dancing relieved the anxiety people felt.

The dancing and singing continued long after the plague was over, but no longer did people of all ages participate. The dance developed jovial aspects and was taken up by children. They also danced in circles, singing the same words, but with a rhyme. The essential meaning of the song has been forgotten as younger children transformed it into what is now a dancing game. They could not understand that they are describing a disease that devastated populations of their ancestors and rewrote the history of Europe.

Thus, in all innocence, children have for centuries sung a nursery rhyme with strangely sinister undertones. Although some of the words have changed through the centuries, the "Ring around the rosy" rhyme was, and is, a vile parody of the Black Death of fourteenth-century England.

"Ring-a-ring o' roses" referred to the red rashlike areas on people who were affected; "a pocket full of posies" alluded to the fact that people believed the evil smells associated with the plague were the poisonous breath of demons who were afflicting the people with the disease. Therefore, the smell of herbs or flowers would ward off the evil smell of death and demons.

"A-tishoo! A-tishoo!" addressed the constant sneezing of plague victims, a common symptom. This line, now sung "ashes, ashes," suggests the final demise, or "ashes to ashes." The concluding line, "We all fall down," refers to what millions of people did. They fell down dead!

As nearly as the original rhyme can be reconstructed, it went as follows:

Ring-a-ring o' roses,
A pocket full of posies,
A-tishoo! A-tishoo!
We all fall down.

With the modern modifications it is very probable that many readers, as children, sang and danced in a circle to "Ring Around the Rosy." Both of the writers of this book did.

• •

"This Is the Forest Primeval"

"King of all the conifers of the world, the noblest of a noble race."

—John Muir

The world of the plants has a king. It is among the oldest and mightiest of living things. It is the giant sequoia or California big tree. Not even in past geologic periods were there trees greater than *Sequoiadendron giganteum*. Those who know the species best maintain that it never dies of disease or senility. If the tree survives predators during its infancy and the hazards of fire in youth, not even a bolt from heaven can end its centuries of life.

The giant sequoias are not only among the largest of trees; they are also among the most limited in range and number of individuals of any major tree species. Native only to central California, on the westerns slopes of the Sierra Nevada, they occupy a total area of less than 1,500 square miles and are restricted to elevations from 4,000 to 8,000 feet above sea level. They occur in a narrow discontinuous belt extending north to south for a distance of about 250 miles.

Although the giant sequoias have many unusual and distinctive characteristics, nothing about them is so spectacular as their size and bulk. Full-grown specimens average 275 feet in height and over 25 feet in diameter. Most of them occur in groves in Sequoia, Kings Canyon, and Yosemite national parks in California.

About 3,500 years ago, at a time when the earliest pharaohs were ruling Egypt, a tiny seed fluttered to a bare patch of ground in what is currently part of Sequoia National Park. It germinated, took root, and flourished, producing the largest tree alive in the modern world. It became the tree now known as the General Sherman.

On August 7, 1879, John Wolverton, a trapper, was at work in the Giant Forest area when he came upon this gigantic sequoia. It was by far the

largest he had yet seen. So impressed was he by its immensity that he named it for the commanding officer under whom he had served during the Civil War. The tree was then and thereafter called the General Sherman tree.

On August 9, 1931, the headline in the *Fresno Republic* announced "Sherman Tree Found Largest of Sequoia Giants." This was the result of weeks of measurements by a number of engineers employed by the Fresno County Chamber of Commerce. They measured several of the largest sequoias in the parks, and the results confirmed the General Sherman as the largest tree in the world. A close runner-up was the General Grant tree, which in some measurements is larger than the king, but total figures showed the General Sherman to be 57,336 board feet larger than its nearest competitor.

The size difference can be better appreciated if it is understood that a fair-sized pine or fir, three feet in diameter and 200 feet tall, might contain as many as 5,000 board feet. This means that the difference in wood content between the General Sherman and the General Grant is the equivalent of nine or ten average-sized pine trees. The General Sherman is estimated to contain 600,120 board feet of timber, the amount of wood needed to make over five billion wooden kitchen matches. Or, put to other uses, its lumber would be sufficient to build forty-five family homes. The same amount of wood could be found on twenty acres of an average California pine forest. The main trunk alone would fill thirty railroad cars.

The tree stands 280 feet tall and would be significantly taller had it not been topped by natural causes, such as lightning. The diameter, measured at the widely expanding base, is over thirty-six feet, and five feet above ground it is over twenty-five feet. The tree does thin upward, but very slightly. At the height of 120 feet the diameter of the trunk is still eighteen feet, clearly showing that its massive girth is maintained.

The bark of the General Sherman is about two feet thick. The tree's largest branch, which is nearly 150 feet from the ground, is over seven feet in diameter and 154 feet long. If it were standing upright, it would dominate almost any tree in the eastern United States. The weight of the General Sherman is about 2,145 tons. Its age is estimated to be about 3,500 years, and it is still growing. The giant stands guard at the entrance to the finest part of the Giant Forest in Sequoia National Park.

Although the General Sherman is the largest, most massive living thing, there are, in the widely varying kingdom of trees, some that are taller, larger in girth, and older. Its cousins, the coast redwoods, are typically fifty feet taller, and one of the trees in Humboldt County's Redwood Creek Grove is over 366 feet tall, having died back from almost 368 feet in the past twenty-five years. As for girth, the fattest tree is a Montezuma cypress in the state of Oaxaca, Mexico, at 117.6 feet in circumference five feet above the ground. In the 1960s, studies of the bristlecone pine, which grows 10,000 feet above sea level on California's eastern Sierra, revealed trees over 4,000 years in age. The Methuselah, the oldest known living bristlecone, is 4,600 years old. Dendrochronologists consider that the potential life span of a sequoia could be 5,000 years; in 1,500 years we will know.

The ancestors of the big trees have a family history that goes back in geologic time to the Mesozoic, the era of the dinosaurs. Their fossils have been found in widely dispersed localities such as Spitsbergen, central and western Europe, China, Japan, and across the North American continent. Fossil sequoia stumps are present in the petrified forest of Yellowstone National Park in Wyoming and among the stony trees of the Petrified Forest National Park in Arizona. No fossil trees have been found that approach the size of the living sequoia. It is sufficient that they have stood as sentinels connecting the time of dinosaurs with a time when humans can explore the heavens. With our stewardship they can be a part of our future.

Through the ages the sequoias living today dominated the forest scene in the mountains and coast of California, unknown to civilized man. Nations rose and fell, civilization spread from Asia to Europe and westward throughout America, but the sequoias were not discovered until, historically speaking, today.

The coast redwoods, *Sequoia sempervirens*, could hardly have been missed by Sir Francis Drake when he and his men landed near San Francisco Bay in 1579, but no mention of them has been recorded. The first mention of these coastal sequoias was made almost 200 years later, on October 10, 1769, by Fray Juan Crespi, who wrote of "very high trees of red color."

In 1833 Lieutenant Joseph Walker led a party across the Sierra Nevada into California and noted stands of giant trees. Other pioneers who followed also referred to the "giants," but nothing official was recorded until 1852. One spring day of that year a miner pursued a grizzly bear far up into the tall timber. History has preserved the miner's name, A. T. Dowd, and the fact that so astonished was he when he encountered the big trees that he let the bear escape. His fellow miners came to see what had so captivated him, and when they departed they spread the fame of the "mammoth trees," as they were first identified.

The tree Dowd initially discovered was a colossus of the Calaveras Grove. It was promptly cut down by the pioneers of the area, and the stump was made into a dance floor on which thirty couples were able to stomp, trot, and waltz simultaneously to the hit bands of the day. This historic relic still endures, and is known as the "Dancehall stump."

In 1847 Stephan Endlicher, an Austrian botanist, named the coast redwoods *Sequoia sempervirens* after Sequoia, the Cherokee Indian. Most

appropriately were they named for the great Cherokee chief who had created an eighty-six-symbol alphabet for his people, representing every sound in their language. He too was a giant.

Scientists from the United States were not the first to describe the big trees. Specimens of wood, bark, and leaves were sent to Washington, D.C., for study in 1853 by way of the isthmus of Panama but never arrived. Meanwhile an Englishman, William Lobb, returned to England with good samples of the tree. From this material, botanist John Lindley properly described the tree and named it *Wellingtonia gigantea* in honor of the duke of Wellington. In 1854 the botanist Joseph Decaisne recognized the new conifer as a species of sequoia and called it *Sequoia gigantea*. Because the genus *Sequoia* had, fortunately, been recorded before *Wellingtonia*, this name became official. The name *Washingtonia*, submitted by possessive patriots, was also too little, too late.

A third species of sequoia was discovered in China in 1941. This tree, previously known only in fossil specimens, bears a strong resemblance to the coast redwood and has been named the dawn tree, or *Metasequoia*. To further distinguish the big trees from the other two sequoias, the name *Sequoia gigantea* had been changed to its present taxonomic classification, *Sequoiadendron giganteum*. Perhaps this most recent name for the big trees will endure long enough to be accurate for the current publication!

In 1856 Hale Tharp was welcomed into the mountain home of the Potwisha tribe of the Monache Indians. He was the first white man they had ever seen. In return for their friendship he shot game for them. Two years later the chief of the tribe, Chappo, invited Tharp to see for himself the giant trees about which the Indians often spoke. He was led into the Giant Forest. What he felt as he first saw the magnificent stands of enormous trees can only be imagined. He was the first of millions to follow who would stand and look up at the magnificent giants. In 1910, as he reminisced, Tharp remarked that he believed he was the first white man to enter the Sequoia National Park region. This claim has never been challenged, as Dowd's earlier discovery had been in the area of Kings Canyon National Park.

Tharp was followed shortly by John Muir, who in 1873 began an exploration and investigation of the sequoias that has remained unsurpassed. He was instrumental in the establishment of national preserves to protect the sequoias for all time.

The giant trees became almost legendary in the West, while the eastern part of the country remained incredulous. In Kings Canyon lies the Centennial stump. This was the first of several big trees cut down for exhibit in the unbelieving world outside California. A section from this tree was shipped to the Centennial Exposition at Philadelphia in 1876. It was almost completely rejected as nongenuine; in fact it was reported as a "tall tale from California."

The upper portion of the Centennial tree, now hollowed by fire, can be seen today lying amid young sequoias near its stump. The Chicago stump is all that remains of a giant that was felled in 1891 for shipment to the Columbian Exposition of 1892 at Chicago. Even at that late date, when logging was in full swing, the easterners continued to reject what their eyes beheld and called it "the fraud from California."

Today's visitors to Kings Canyon National Park can inspect the Chicago stump and climb a permanent ladder to stand on the remains of what was once a giant tree. From the top the view is of a dead forest of stumps, the legacy left by the fruitless experiment in logging, a monument to man's greed and lack of vision.

The skepticism expressed by easterners was not shared by the timber companies. Between 1862 and 1900 logging operations wiped out much of the finest forest in the world. Those sequoias cut in the Converse Basin are thought by many to have been the most magnificent stand of giant sequoias in existence.

Because the chief virtue of the sequoias is their durability and bulk, they were sought out by lumber companies as soon as their existence was confirmed. Judging by the appearance of the groves these trees promised ready fortunes. The number of board feet of wood produced by such trees could be matched nowhere else in the world, at least in the minds of the lumber executives.

So logging railroads were hurriedly built up the mountains, mills were assembled, and Lilliputian lumberjacks went to work among woody Brobdingnagians. Platforms were erected above the flaring base where two men, standing on each end, cut an enormous wedge at least ten feet into the tree. The fallers then went to the opposite side of the tree and, for days, dragged a twenty-foot saw back and forth, cutting into the heart of the tree. All the while the saw would be greased so it wouldn't stick, and wedges would be driven into the long cut to keep the saw from binding. After almost two weeks of the men working twelve hours a day and six days a week, the first tree started to lean, and to the lusty cry of "Timber!" it began to topple. The tree struck the earth with a crack of limbs and a seismic shock that was felt at least a mile away.

To the consternation of all, dollar signs faded from their dreams of power and glory, because the big trees were so huge and brittle that they shattered on impact with the ground. Usually only 20 to 40 percent of the log was recovered, and the remainder of the tree lay where it fell.

Various methods were applied by the lumberjacks to try to soften the fall, such as "featherbedding." A ditch was dug in the projected line of the fall and filled with branches and leaves to cushion its impact, but it helped very little. Another idea was to fell the tree uphill so it would have a shorter distance to drop, another futile and ineffective procedure.

It is doubtful that any of the logging companies made a profit from the giant sequoias. All a visitor needs to do is visit the Big Stump Basin and view the result of this useless devastation. Along the tourist trail are numerous remains of fallen giants. A century of evidence remains, for even when the trees are destroyed they are slow to decay, showing how little of the actual log was carted off to the lumber mill. An excellent example of waste can be seen in the carcass of the Shattered Giant. The timbering of this sequoia was so poorly done that when it impacted with the earth it fractured so completely that none of it was recovered. The entire ghastly logging enterprise ended in financial disaster, but the ravaging of the forest was an almost complete success.

The long battle to save the big trees began in earnest among concerned Californians in the late nineteenth century. The movement was championed by the influential George Stewart, editor of the *Visalia Delta*. His fiery editorials inspired many people to flock to the cause, for under government land laws the Giant Forest could still revert to the open market and be sold. Not only the timber lords, but damage caused by sheep and man-made fires could destroy the forest area. Concerned citizens from the nearby counties petitioned Congress to establish a national park to protect the entire watershed. Not without resistance, a first step was achieved in 1890, when the Garfield Grove was set aside as a preserve. Enlargement of the park began almost immediately, and today almost all of the big trees are enclosed and protected in Sequoia, Kings Canyon, and Yosemite national parks. With continued vigilance, never again will they be vandalized.

The giant sequoia and its cousin, the coast redwood, are alike in a number of ways. Both are giant trees, the wood is pink or red when freshly cut, and both are quite resistant to decay and attacks from insects or fungus. For many years, unfortunately, the lords of lumber detected another similarity; both species represented a bonanza crop, to be ravaged irresponsibly until they ceased to turn a profit, until the lust to bring down the big ones was satisfied, or until another crop of trees was ready.

Over a hundred years ago the coast redwoods grew in an unbroken belt along the Pacific coast no more than thirty miles wide, from southwestern Oregon to the Big Sur country south of San Francisco. The redwoods grow only as far as morning fogs can roll inward from the coast. In these foggy areas they have always been "Monarchs of the Forest." The Indians who once lived in the redwood forests looked on these great trees as sacred.

The arrival of large numbers of settlers in the late nineteenth century inaugurated a period of reckless exploitation. As a result the original two million acres of redwood forest was reduced to less than 300,000 acres by 1965. Most of the trees tumbled to the music of chain saws, draglines, and tractors.

Coast redwoods can be logged when they are less than one hundred years old and can be replaced provided that the land favorable to their growth is not laid waste. After the supertractor and power saw were put on the job, reforestation could not compete with the tempo of felling. To preserve the forest grandeur of old-growth redwoods, which had taken a thousand years to produce, the Save-the-Redwoods League was formed. As a result of its efforts President Lyndon Johnson signed a law creating the Redwood National Park in 1968. It was enlarged by President Carter in 1978, giving protection and preservation to about 106,000 acres of coast redwoods. Logging of the redwoods outside the park area still continues, and the park service estimates that by the year 2000 practically no groves of redwood will exist outside the park boundaries.

The coast redwoods cannot compare with their cousins, the big trees, in longevity, but they have a fairly long fling at living. The normal life expectancy of the redwood is 1,000 to 1,500 years, although many live longer. The oldest specimen whose rings have been counted is estimated to have been about 2,200 years old. This tree began to grow when Hannibal was taking his elephants over the Alps and was already about 200 years old when Christ was born. A section of this log has been preserved in the national park.

Despite their similarities, the coast redwood and giant sequoia differ vastly in appearance, habitat, growth patterns, and reproduction. They also differ in foliage cones, bark, and root structure. Each has a specific set of climatic restrictions and will tolerate only slight variations. Sequoia bark reaches a thickness of two feet, twice that of the redwood, and its cones, though small, are twice the size of its coastal cousins.

The two trees appear more similar in youth, growing straight with conical outlines like a perfect Christmas tree. Then, well after its hundredth year, the sequoia begins to mature; it increases in breadth, and great

REDWOOD

GIANT SEQUOIA

arms begin to appear, replacing the droopy boughs of youth. Very little is added to its height, but an ordinary big tree can add several thousand board feet per year, mostly in girth. The enormous superstructure is buttressed by a gigantic base and root system. Although shallow, seldom deeper than three feet, the roots spread out in a broad circle covering an acre or more.

Both of the trees put up a remarkable fight for survival, the giant sequoia by way of self-preservation and the redwood by its remarkable reproductive alternatives. When cut down, the coast redwood will sprout readily from the stump, throwing up a ring or circle of young trees. A fallen log may send up shoots, the result being a straight row of close-ranked trees that put down roots straddling the log. According to foresters these sprouts grow so rapidly that in many cut-over areas they may be large enough to harvest again within forty to fifty years. The sprouts that grow from the stumps are really sister trees. They are not offspring of the old tree, but a continuation of the same life, adding to the previous tree in an expression of immortality. A new generation of trees is produced only from the seed.

The big tree, *Sequoiadendron giganteum*, reproduces only from seed. This in itself is spectacular. The cone, usually about three inches long, contains several hundred tiny seeds, each the size of a pinhead and weighing about .00016 ounce (6,000 = one ounce). The seeds resemble small flakes of breakfast cereal such as uncooked oatmeal. Only the thin, dark line in the center of each seed contains the life germ, the embryonic giant tree. It is difficult to realize that from this almost microscopic seed will grow a tree twenty-five feet in diameter and weighing thousands of tons.

The remainder of the seed is a pair of golden wings that will carry the embryo through the air to the forest floor. Annually the trees will rain down millions of seeds, but with the heavy shade cast by the parent tree, the matted roots beneath them, and the floor covering of forest debris, the seeds have little chance of germinating. In fact scientists rate the odds against a single seed becoming a mature tree at about a billion to one.

Of the millions of seeds that float down to the forest floor annually, perhaps one will find conditions suitable for germination. The seed must fall on a rich mineral soil from which decaying leaves, branches, and other forest duff have been removed by fire, erosion, or other causes. Along with proper soil, plenteous sun and moisture are required. When perfect conditions are discovered, many seeds will germinate, and most seedlings will become a meal for wood ants, squirrels, chipmunks, deer, finches, sparrows, and other creatures of the food web. Some seedlings will make it through the first year and will face the next few centuries along with others that found the same ideal conditions. This is why the big trees are typically found in clusters, or groves, where trees of uniform size are separated some distance from other giant sequoias.

One of the important requirements for the germination of the giants is fire. They are, in fact, so fire dependent that without the occasional intrusion of fire the giant sequoia would not persist long on this earth. Adequate moisture is, of course, necessary for the seed to sprout, but this is rarely a problem at elevations above 4,000 feet. The seed must also have exposed mineral-rich soil and direct sunlight, which is where fire becomes so important. Fire prepares the seedbed by consuming forest litter, making it possible for the seed to reach a soft mineral soil. It also creates openings in the forest canopy, thereby allowing sunlight, another basic ingredient, to reach the sequoia seedlings. While fire, usually from lightning, is at work it also removes the competition of other trees and returns nutrients, tied up in forest litter, to the soil, where they are again available for sprouting seeds. Fire may even be credited with producing hot air that causes the cones to dry out and open, for the seeds might otherwise remain enconed for twenty years or more.

When a giant sequoia falls naturally, its head is smashed into fragments that are eventually consumed by a hunting fire, while the trunk is slowly wasted away by centuries of fire and weather. One of the most interesting fire activities on the fallen trunk is the boring of great tunnellike hollows, while the bark, which contains no resin and is fire-resistant, usually remains intact. All of the famous hollow trees were excavated by fire, for the *Sequoiadendron* is rarely hollowed by decay.

The best example of a prostrate giant hollowed by fire is the Fallen Monarch, located in Kings Canyon National Park. This is the tree through which, years ago, a person could ride on horseback. Although gradual subsidence has lessened the overhead clearance, anyone under six feet can still walk through the fallen tree without stooping. In 1868 one of the first settlers in the area lived in this log for four years until his cabin was built.

Folklore about the Fallen Monarch recounts that shepherds and cattlemen used the hollow as headquarters and that in lumbering days a saloon was maintained within the tree. Prior to 1913 the United States Cavalry patrolling the park area used the hollow log as a stable.

The big tree keeps its youth far longer than any of its neighbors. Most firs are old in their second or third century of life, pines become aged in their fourth or fifth, while the giant sequoia growing beside them is still in the bloom of youth. It is juvenile in every feature while the neighboring pines are in their old age. It can be said that these giants will be nowhere near their prime size or beauty before their 1,500th birthday and, under favorable conditions, will not be old before their 3,000th year. Many, no doubt, will reach a riper old age than that.

The age of the trees has been determined through growth ring counts. Each year that the tree lives it adds an additional layer of wood and bark. If the tree is cut down, the rings of each year's growth, visible even to an untrained eye, can be counted accurately by anyone with adequate patience. Since growth takes place only in the cambium, the narrow zone between the sapwood and the bark, the outermost ring represents the last year of its life. A sufficient number of growth rings were exposed during the days of exploitation to determine the ages of many of the giants. The ages of several were determined to be over 3,000 years, and the Chicago stump, mentioned earlier, still stands to remind us that nothing has been added to its ring count of 3,126 years since the lumber lords cut it down.

In his writings John Muir claims to have counted over 4,000 rings in one tree, but the location of the "Muir stump" is unknown. Quite possibly Muir was correct in his count, and the stump may have been burned and the ring count destroyed by the many fires that swept the lumbered areas. Because this count is based on a missing stump, it has not been recognized.

The relative size and age of the *Sequoiadendron* are determined by the conditions of the soil, water supply, and exposure to the sun. Therefore the famous Dancehall stump, discussed earlier, is twenty-six feet in diameter at the cut, yet a tree ring count showed this tree to be only 1,200 years old when felled. When standing, it easily rivaled the biggest and oldest trees in size and bulk despite its relative youth.

Many botanists believe that the big trees, if undisturbed by nature or man, could live to be 4,000 years old and that a few might even reach 5,000 years of age. After 3,000 years the tree is ripe and begins to decay in the heart, but the vitality of this species is so great that many centuries pass before it actually dies.

The reason for the trees' longevity might be, as George Stewart observed, that "the giant sequoias never learned how to die." Their long lease on life is due largely to the extraordinary quality of the bark, which in some cases is two feet thick. The bark is highly flavored with tannin, which makes the tree almost impervious to attack by any known species of insect. The thick bark contains no resin, the flammable substance that makes many types of bark burn so readily. It is spongy and fibrous and is just about as fireproof as asbestos. There is a recorded incident of a July storm in which the top of one giant tree was struck by a lightning bolt. It smoldered quietly without much damage to the tree until it was put out by a snowstorm that occurred the following October.

The big trees' reaction to fire damage or other threats to life and limb is to start healing. Growth over an injured area is twice as fast as normal, so that a three-foot scar can be healed in roughly one hundred years. The Sawed Tree in Big Stump Basin was destined for the lumber mill in the last century. After a wedge had been cut partway through the trunk, lumbering was halted, no doubt because of a change of mind, heart, or economics. The tree is still covering its scars and is being nourished by the uncut area.

The national parks evermore protect the virgin growth of the giant sequoias, and in several more centuries we will discover just how long the big tree can live. There are at present about 20,000 trees larger than ten feet in diameter. In the meantime several groves of younger sequoias are on their way. In Crescent Meadow there is a grove of 500-year-old juveniles. In the Big Stump Basin some enterprising loggers, with an eye to the future of the industry, planted a group of trees that have just passed the century mark. Along the Generals Highway a number of young trees sprang up after road construction in 1929 and are now about fifty feet high. It is the privilege of any generation of humans to observe about one-fiftieth of a big tree's life, from seeding to sapling to patriarch, while it stands as a sentinel guarding our once and future history.

The National Park Service has established a number of self-guided nature trails, some of which go well into the backcountry. Those who take the longest paths are quickly separated from the impact of civilization. Here it becomes no strain on the imagination, when encountering the groves of giants, to feel as if one is peering into the geologic past. A *Triceratops* might be rubbing its horns against a hearty ancestral sequoia, while overhead *Pterodactyls* flap from tree to tree searching for insects. In truth the forest of the giants was, as it is today, the forest primeval.

• •

BIBLIOGRAPHY

Books

Adams, Kramer. *The Redwoods*. New York: Popular Library, no date.
Andrews, Roy Chapman. *Meet Your Ancestors*. New York: Viking Press, 1945.
———. *All About Strange Beasts of the Past*. New York: Random House, 1956.
Ardrey, Robert. *African Genesis*. New York: Delta, 1961.
Asimov, Isaac. *The Egyptians*. Boston: Houghton Mifflin, 1967.
Attenborough, David. *The First Eden*. Boston: Little Brown, 1987.
———. *Life on Earth*. Boston: Little Brown, 1987.
———. *The Living Planet*. Boston: Little Brown, 1984.
Baker, Robin, ed. *The Mystery of Migration*. New York: Viking Press, 1981.
Bakker, Robert T., Ph.D. *The Dinosaur Heresies*. New York: Morrow, 1986.
Bardach, John. *Harvest of the Seas*. New York: Harper & Row, 1968.
Barnyard, P. J. *Natural Wonders of the World*. Secaucus, N.J.: Chartwell Books, 1978.
Blond, Georges. *Great Migrations*. London: Hutchinson, 1958.
Bok, Bart, and Priscilla Bok. *The Milky Way*. Cambridge, Mass.: Harvard University Press, 1981.
Bordes, François. *The Old Stone Age*. New York: McGraw-Hill, 1968.
Brackman, Arnold C. *The Search for the Gold of Tutankhamen*. New York: Mason/Charter, 1976.
Bronowski, J. *The Ascent of Man*. Boston: Little Brown, 1974.
Burland, C.A. *Montezuma*. New York: Putnam, 1975.
Burton, Maurice, & Robert Burton. *Inside the Animal World*. New York: Quadrangle, 1977.
———. *Encyclopedia of Fish*. New York: Finsburg Books, 1984.
Burton, Robert. *The Life and Death of Whales*. New York: Universe Books, 1980.
Calder, Nigel. *The Restless Earth*. New York: Viking Press, 1972.
———. *The Comet Is Coming*. New York: Viking Press, 1981.
———. *Timescale*. New York: Viking Press, 1983.
Calvocoressi, Peter. *Who's Who in the Bible*. New York: Viking Press, 1987.
Caras, Roger. *Dangerous to Man*. New York: Stoeger Publishing Co., 1974.
Carr, Donald E. *The Deadly Feast of Life*. New York: Doubleday, 1971.

Carrington, Richard. *Mermaids and Mastodons.* New York: Rinehart and Company, 1957.
Ceram, C. W. *Gods, Graves & Scholars.* New York: Knopf, 1959.
———. *Hands on the Past.* New York: Knopf, 1966.
———. *The First American.* New York: Mentor, 1971.
Chapman, Clark R. *The Inner Planets.* New York: Charles Scribner's Sons, 1977.
Chard, Chester S. *Man in Prehistory.* New York: McGraw-Hill, 1969.
Chinery, Michael. *Killers of the Wild.* New York: Chartwell Books, 1979.
Chorlton, Windsor. *Ice Ages.* New York: Time-Life Books, 1983.
Clapham, Francis, ed. *Our Human Ancestors.* New York: Warwich Press, 1976.
Clark, Grahame. *World Prehistory.* London: Cambridge, 1977.
Clausen, Lucy W., Ph.D. *Insect Fact and Folklore.* New York: Macmillan, 1954.
Cloud, Preston. *Cosmos, Earth and Man.* New Haven: Yale University Press, 1978.
———. *Oasis in Space.* New York: Norton, 1988.
Colbert, Edwin H. *Men and Dinosaurs.* New York: E. P. Dutton, 1968.
———. *Wandering Lands and Animals.* New York: E. P. Dutton, 1973.
———. *Dinosaurs, an Illustrated History.* Maplewood, N.J.: Hammond, 1983.
Constable, George. *The Neanderthals.* New York: Time-Life Books, 1973.
Constance, Arthur. *The Impenetrable Sea.* New York: The Citadel Press, 1958.
Cornell, James C., and E. Nelson Hayes, eds. *Man and Cosmos.* New York: W. W. Norton, 1975.
Cornwall, I. W. *The World of Ancient Man.* New York: John Day Co., 1964.
Cousteau, Captain J. Y., with James Dugan. *The Living Sea.* New York: Harper & Row, 1963.
Daniels, George G., ed. *Volcano.* New York: Time-Life Books, 1982.
de Camp, L. Sprague, and Catherine Crook de Camp. *The Day of the Dinosaur.* New York: Doubleday, 1968.
De Nevi, Don. *Earthquakes.* Berkeley, Calif.: Celestial Arts, 1977.
Derry, François. *Our Unknown Earth.* New York: Stein and Day, 1967.
Desmond, Adrian J. *The Hot Blooded Dinosaurs.* New York: Dial Press, 1976.
Dickinsen, Joan Younger. *The Book of Diamonds.* New York: Crown, 1965.
Dickinson, Terence. *The Universe Beyond.* Ontario, Canada: Camden House, 1986.
Dineley, David. *Earth's Voyage Through Time.* New York: Knopf, 1974.
Douglas-Hamilton, Iain, and Oria Douglas-Hamilton. *Among the Elephants.* New York: Viking Press, 1975.
Dröscher, Vitus B. *The Friendly Beast.* New York: E. P. Dutton, 1971.
———. *They Love and Kill.* New York: E. P. Dutton, 1976.
Dryson, James L. *The World of Ice.* New York: Knopf, 1962.
Edey, Martland. *The Missing Link.* New York: Time-Life Books, 1972.
Ehrlich, Paul, and Anne Ehrlich. *Extinction.* New York: Random House, 1981.
Ericsen, David B., and Goesta Wollen. *The Ever Changing Sea.* New York: Knopf, 1967.
Evans, Howard E. *Life on a Little Known Planet.* New York: E. P. Dutton, 1968.
Fabre, J. Henri. *The Life of the Spider.* New York: Dodd, Mead, & Co., 1919.

Fagan, Brian M. *The Rape of the Nile.* New York: Charles Scribner's Sons, 1975.
———. *The Great Journey.* London: Thames and Hudson, 1987.
Fairservis, Walter A., Jr. *The Threshold of Civilization.* New York: Charles Scribner's Sons, 1975.
Fisher, Helen E. *The Sex Contract.* New York: Quill, 1983.
Foreman, Grant. *The Five Civilized Tribes.* Norman, Okla.: University of Oklahoma Press, 1977.
Frank, Claus Jurgen, ed. *Wonders of Nature.* New York: Macmillan, 1980.
Frazier, Kendrich. *The Violent Face of Nature.* New York: Morrow, 1979.
Frisch, Karl Von. *Animal Architecture.* New York: Harcourt Brace Jovanovich, 1974.
Fuchs, Sir Vivian, ed. *Forces of Nature.* New York: Holt, Rinehart and Winston, 1977.
Fuller, Errol. *Extinct Birds.* New York: Facts on File Publications, 1987.
Gardner, Martin. *Fads & Fallacies.* New York: Dover, 1957.
George, Uwe. *In the Deserts of This Earth.* New York: Harcourt Brace Jovanovich, 1976.
Goodall, Hugo, and Jan van Lawich. *Innocent Killers.* New York: Ballantine Books, 1970.
Goodwin, John. *This Baffling World.* New York: Hart, 1968.
Grant, Michael. *The Rise of the Greeks.* New York: Charles Scribner's Sons, 1987.
Gribben, John. *This Shaking Earth.* New York: Putnam, 1978.
Grimm, William C. *Familiar Trees of America.* New York: Harper & Row, 1967.
Gropman, Donald, with Mirus Kenneth. *Comet Fever.* New York: Fireside Book, 1985.
Grossman, Louise, Shelly Grossman, and John Hamlet. *Our Vanishing Wilderness.* New York: Grosset & Dunlap, 1972.
Grzimek, H. C. Bernhard, ed. *Animal Life Encyclopedia.* Vols. 1-13. New York: Van Nostrand Reinhold, 1984.
Hadas, Moses. *Imperial Rome.* New York: Time-Life Books, 1965.
Hadingham, Evan. *Circles and Standing Stones.* New York: Walker and Co., 1975.
———. *Secrets of the Ice Age.* New York: Walker and Co., 1979.
Hallam, A. *A Revolution in Earth Science.* New York: Clarendon Press, 1973.
Harris, S. L. *Fire and Ice.* Seattle, Wash.: Pacific Search Press, 1980.
Hawkings, Jacquetta. *The Atlas of Early Man.* New York: St. Martin's Press, 1976.
Hayes, Harold T. P. *The Last Place on Earth.* New York: Stein and Day, 1977.
Heilbrin, Angelo. *Mont Pelée and the Tragedy of Martinique.* Philadelphia: J. P. Lippincott, 1905.
Helm, Thomas. *Hurricanes: Weather at Its Worst.* New York: Dodd, Mead & Co., 1967.
Hemming, John. *The Search for El Dorado.* New York: E. P. Dutton, 1978.
Herm, Gerhard. *The Phoenicians.* New York: Morrow, 1975.
Hoffman, Michael A. *Egypt Before the Pharaohs.* New York: Knopf, 1979.

Hohn, Reinhardt. *Curiosities of the Plant Kingdom*. New York: Universe Books, 1980.
Holmes, William D. *Safari*. New York: Coward McCann, Inc., 1960.
Hopper, R. J. *The Early Greeks*. New York: Barnes & Noble, 1976.
Howells, William. *Mankind in the Making*. New York: Doubleday, 1967.
Hoyle, Fred. *Galaxies, Nuclei and Quasars*. New York: Harper & Row, 1965.
Hunt, John. *A World Full of Animals*. New York: David McKay, 1969.
Iacopi, Robert. *Earthquake Country*. Menlo Park, Calif.: A Sunset Book, 1976.
Janus, Christopher G., with William Brashler. *The Search for Peking Man*. New York: Macmillan, 1975.
Jenkins, Alan C. *Mysteries of Nature*. New York: Facts on File, 1984.
Jenkinson, Michael. *Beasts Beyond the Fire*. New York: E. P. Dutton.
Johanson, Donald, and Maitland Edey. *Lucy*. New York: Simon and Schuster, 1981.
Kovalik, Vladimir, and Nada Kovalik. *The Ocean World*. New York: Holiday House, 1966.
Kroeber, Theodora. *Ishi*. Berkeley, Calif.: University of California Press, 1973.
Krupp, E. C. *Echoes of the Ancient Skies*. New York: Harper & Row, 1983.
Kurtén, Björn. *The Cave Bear Story*. New York: Columbia University Press, 1976.
Lambert, David. *Dinosaurs*. New York: Crown, 1978.
Lane, Frank W. *Animal Wonder World*. New York: Sheridan House, 1951.
———. *The Elements Rage*. New York: Chilton Books, 1965.
———. *The Violent Earth*. Topsfield, Mass.: Salem House, 1986.
Lavine, Sigmund. *Strange Travelers*. Boston: Little Brown, 1960.
Leakey, Mary. *Disclosing the Past*. New York: Doubleday, 1984.
Leakey, Richard E. *The Making of Mankind*. New York: E. P. Dutton, 1981.
Leakey, Richard, and Roger Lewin. *People of the Lake*. New York: Doubleday, 1978.
Levin, Harold L. *The Earth Through Time*. Philadelphia: Sanders College Publishers, 1983.
Lewin, Roger. *Thread of Life*. Washington, D.C.: Smithsonian Books, 1982.
Lopez, Barry Holstun. *Of Wolves and Men*. New York: Charles Scribner's Sons, 1978.
Lucas, Frederic A. *Animals of the Past*. New York: American Museum of Natural History, 1929.
MacGregor, Alasdair Alpin. *The Changing Land*. Bath, England: Kingsmead Press, 1973.
Mackal, Roy P. *Searching for Hidden Animals*. New York: Doubleday, 1980.
Maffei, Paolo. *Monsters in the Sky*. Cambridge: M.I.T. Press, 1976.
Manning, Reg. *What Kinda Cactus Izzat?* Phoenix, Ariz.: Reganson Cartoon Books, 1941.
Mason, George F. *Animal Vision*. New York: Morrow, 1968.
Martin, John M., and Michael Martin. *The Book of Beasts*. New York: Viking Press, 1981.
Marx, Robert. *Sea Fever*. New York: Doubleday, 1972.
McClung, Robert M. *Lost Wild America*. New York: Morrow, 1960.

McCormick, Harold W., Tom Allen, and William Young. *Shadows in the Seas.* New York: Weathervane Books, 1968.
McCracken, Harold. *The Charles M. Russell Book.* New York: Doubleday, 1957.
McNeill, William H. *Plagues and Peoples.* New York: Doubleday, 1976.
McNulty, Faith. *Must They Die? The Strange Case of the Prairie Dog and the Black-Footed Ferret.* New York: Doubleday, 1971.
Miller, Robert C. *The Sea.* New York: Random House, 1966.
Milne, Lorus, and Margery Milne. *The World of Night.* New York: Harper & Brothers, 1956.
Minton, Sherman A., and Madge R. Minton. *Venomous Reptiles.* New York: Charles Scribner's Sons, 1969.
Montagu, Ashley. *Man: His First Two Million Years.* New York: Delta Books, 1969.
Moore, Patrick. *Armchair Astronomy.* New York: W. W. Norton, 1984.
Morris, Desmond. *The Naked Ape.* New York: Dell, 1969.
———. *Catlore.* New York: Crown, 1987.
Morrison, Philip, and Phylis Morrison. *The Ring of Truth.* New York: Random House, 1987.
Motz, Lloyd, Ph.D., ed. *Rediscovery of the Earth.* New York: Van Nostrand Reinhold, 1975.
Myles, Douglas. *The Great Waves.* New York: McGraw-Hill, 1985.
O'Neil, Paul. *Gemstones.* New York: Time-Life Books, 1983.
Panati, Charles. *Extraordinary Origins of Everyday Things.* New York: Harper & Row, 1987.
Paul, Gregory S. *Predatory Dinosaurs of the World.* New York: Simon & Schuster, 1988.
Peattie, Donald Culross. *A Natural History of Western Trees.* Boston: Houghton Mifflin, 1953.
Pfeiffer, John E. *The Emergence of Man.* New York: Harper & Row, 1972.
Preiss, Byron, ed. *The Universe.* New York: Bantam, 1987.
Preston, Douglas., Jr. *Dinosaurs in the Attic.* New York: St. Martin's Press, 1980.
Prince, J. H. *Languages of the Animal World.* New York: Thomas Nelson, 1975.
Pyke, Magnus. *The Delights of Science.* New York: Sterling, 1977.
Quammen, David. *The Flight of the Iguana.* New York: Delacorte Press, 1988.
———. *Natural Acts.* New York: Laurel, 1982.
Raup, David M. *The Nemesis Affair.* New York: W. W. Norton, 1986.
Reader, John. *Missing Links.* Boston: Little Brown, 1981.
Redfern, Ron. *The Making of a Continent.* New York: Times Books, 1983.
Ricciuti, Edward. *Killers of the Sea.* New York: Macmillan, 1973.
———. *Killer Animals.* New York: Walker and Co., 1976.
Ritchie, David. *The Ring of Fire.* New York: Mentor, 1982.
Ronan, Colin. *Lost Discoveries.* New York: Bonanza Books, 1976.
Rowen, Robinson. *Cosmic Landscape.* Oxford, England: Oxford University Press, 1979.
Rudwich, Martin J. S. *The Meaning of Fossils.* New York: Science History Pub., 1976.

Sanderson, Ivan T. *Investigating the Unexplained.* Englewood Cliffs, N.J.: Prentice-Hall, 1972.
Sarton, George. *A History of Science.* Cambridge: Harvard University Press, 1953.
Scheffer, Victor. *The Year of the Whale.* New York: Charles Scribner's Sons, 1969.
Shapiro, Harry L. *Peking Man.* New York: Simon and Schuster, 1974.
Sheehan, Angela, ed. *The Prehistoric World.* New York: Warwich, 1975.
Sherratt, Andrew, ed. *The Cambridge Encyclopedia of Archeology.* New York: Crown, 1980.
Short, Nicholas M. *Planetary Geology.* Englewood Cliffs, N.J.: Prentice-Hall, 1975.
Shreeve, James. *Nature, the Other Earthlings.* New York: Macmillan, 1987.
Sieveking, Ann. *The Cave Artists.* London: Thames and Hudson, 1979.
Silverberg, Robert. *Prehistoric Man in Europe.* New York: New York Graphic Society, 1967.
———. *Mammoths, Mastodons and Man.* New York: McGraw-Hill, 1970.
Smith, Howard E. *Killer Weather.* New York: Dodd, Mead & Co., 1982.
Snow, Dean. *The Archeology of North America.* New York: Viking Press, 1976.
Soustelle, Jacques. *Daily Life of the Aztecs.* New York: Macmillan, 1962.
Sparks, John. *The Discovery of Animal Behavior.* Boston: Little Brown, 1982.
St. John, Jeffrey. *Noble Metals.* Alexandria, Va.: Time-Life Books, 1984.
Stierlin, Henri. *The World of the Pharaohs.* New York: Sunflower Books, 1978.
Stommel, Henry, and Elizabeth Stommel. *Volcano Weather.* Newport, R.I.: Seven Seas Press, 1983.
Strong, Douglas H. *Trees . . . or Timber?* Three Rivers, Calif.: Sequoia Natural History Association, 1986.
Sullivan, Walter. *Continents in Motion.* New York: McGraw-Hill, 1974.
———. *Landprints.* New York: New York Times Books, 1984.
Sunset Books. *National Parks of the West.* Menlo Park, Calif.: Lane Publishing, 1980.
Sutton, Ann, and Myron Sutton. *Nature on the Rampage.* New York: J. P. Lippincott, 1962.
Tarbuck, Joseph J., and Frederick K. Lutgens. *The Earth.* Columbus, Ohio: Merrill, 1984.
Teale, Edwin Way. *The Wilderness World of John Muir.* Boston: Houghton Mifflin, 1954.
Ternes, Alan, ed. *Ants, Indians and Little Dinosaurs.* New York: Charles Scribner's Sons, 1975.
Thomas, Gorden, and Max M. Witts. *The Day the World Ended.* New York: Ballantine Books, 1969.
Thomas, Lewis. *The Medusa and the Snail.* New York: Viking Press, 1979.
Thomas, Lowell. *Seven Wonders of the Ancient World.* New York: Hanover House, 1956.
Thompkins, Peter. *Secrets of the Great Pyramid.* New York: Harper & Row, 1971.
———. *Mysteries of the Mexican Pyramids.* New York: Harper & Row, 1976.

Thorndike, Joseph J., ed. *Mysteries of the Deep*. New York: American Heritage, 1980.
Tilden, Freeman. *The National Parks*. New York: Knopf, 1968.
Time-Life Books, ed. *Voyage Through the Universe: Stars*. Alexandria, Va.: Time-Life Books, 1989.
Vine, Louis L., D.V.M. *Dogs, Devils & Demons*. New York: Exposition Press, 1971.
———. *Your Neurotic Dog*. New York: Dial Press, 1983.
Von Hagen, Victor W. *Realm of the Incas*. New York: Mentor Books, 1963.
Waechter, John. *Man Before History*. New York: E. P. Dutton, 1976.
———. *Prehistoric Man*. London: Octopus, 1977.
Walker, Charles. *Wonders of the Ancient World*. New York: Crescent, 1980.
Waltham, Tony. *Catastrophe, the Violent Earth*. New York: Crown, 1978.
Weiner, Jonathan. *Planet Earth*. New York: Bantam, 1986.
Weisberg, Joseph S. *Meteorology*. Boston: Houghton Mifflin, 1981.
Wendt, Herbert. *In Search of Adam*. Cambridge: Riverside Press, 1956.
———. *From Ape to Adam*. New York: Bobbs-Merrill Co., 1972.
White, Edmund, and Dale Brown. *The First Men*. New York: Time-Life Books, 1973.
White, J. F. *Study of the Earth*. New York: Prentice-Hall, 1962.
White, John R., and Samuel J. Pusateri. *Sequoia and Kings Canyon National Parks*. Stanford, Calif.: Stanford University Press, 1949.
Wilcoxson, Kent H. *Chains of Fire: The Story of Volcanos*. New York: Chilton, 1966.
Wilford, John Noble. *The Riddle of the Dinosaur*. New York: Knopf, 1985.
Woodbury, David. *The Great White Mantle*. New York: Viking Press, 1962.
Woodward, Ian. *The Werewolf Delusion*. New York: Paddington Press, 1979.
Wylie, Francis E. *Tides*. Brattleboro, Vt.: Stephen Greene Press, 1979.

Periodical References

———. "Celebration of a Volcano." *Science News*. January 24, 1987.
———. "Evolving Views of Dinosaurs." *Natural History*. December 1987.
———. "Fertile Ferret?" *Natural History*. May 1987.
———. "Flying Giants Found (*Argentavis magnificens*)." *Bioscience*. December 1980.
———. "Halley's Tailings." *Science 85*. March 1985.
———. Postscript: "Baby Boom." *Natural History*. August 1988.
———. Update: "Dinosaur Ballast." *Science Digest*. December 1981.
———. Update: "The Plight of the Black-Footed Ferret." *Discover*. February 1986.
———. Up Front: "A Shocking Tale About Dummies That Smart." *Discover*. July 1986.
———. "Whale and Farewell." *Discover*. May 1985.
Bartusiak, Marcia. "Secrets of the Glaciers." *Discover*. January 1981.
Bird, Junius. "Legacy of the Stingless Bee." *Natural History*. November 1979.

Bond, Constance. "The Swift Spider That Is Nature's Smallest 'Angler.' " *Smithsonian*. July 1980.
Bower, Bruce. "Ancient Human Ancestors Got All Fired Up." *Science News*. December 10, 1988.
———. "Fossil Find May Be Earliest Known Hominid." *Science News*. April 14, 1984.
———. "Rivers in the Sand." *Science News*. August 26, 1989.
Bowser, Hal. "Meat-Eating Plants." *Science Digest*. November 1981.
Brower, Lincoln P. "Monarch Migration." *Natural History*. June-July 1977.
Brian, C. K. "A Hominid Skull's Revealing Holes." *Natural History*. December 1974.
Burner, David A. "Life on the Cheetah Circuit." *Natural History*. May 1982.
Casson, Lionel. "Think That Taxes Take a Big Bite Today?" *Smithsonian*. March 1989.
Conniff, Richard. "Mice Agree: East or West, Home Is Best." *Smithsonian*. October 1988.
Cook, Robert Edward. "Long-Lived Seeds." *Natural History*. February 1979.
Frost, Honor. "How Carthage Lost the Sea." *Natural History*. December 1987.
Garrett, Wilbur E. "Where Did We Come From?" *National Geographic*. October 1988.
Geist, Valerius. "A Naturalist at Large." *Natural History*. January 1981.
Gilbert, Bill. "Why Don't We Pull the Plug on the Condor and Ferret?" *Discover*. July 1986.
Gorman, James. "Elephant Watching." *Discover*. April 1981.
———. "The Making of a Fossil." *Discover*. January 1981.
Gould, James L. "Do Honeybees Know What They Are Doing?" *Natural History*. June-July 1979.
Gould, Stephen Jay. "The Lesson of the Dinosaurs." *Discover*. March 1987.
Hall, Stephen S. "The Invader." *Hippocrates*. September-October 1987.
Hallett, John. "When Hail Breaks Loose." *Natural History*. June 1980.
Hemming, Christopher F. "A New Plague of Locusts." *Natural History*. December 1978.
Herbert, Wray. "Lucy's Family Problems." *Science News*. July 2, 1983.
———. "Lucy's Uncommon Forebear." *Science News*. February 5, 1983.
Jackson, Peter P. R. "Scientists Hunt the Bengal Tiger—But Only in Order to Trace and Save It." *Smithsonian*. August 1978.
Jones, Robert A. "After a Long and Bitter Battle, a Clear Victory for the Redwoods." *Smithsonian*. July 1978.
Kimeu, Kamoya. "Adventures in the Bone Trade." *Science 86*. March 1986.
Lansford, Henry. "The Frightening Mystery of the Electrical Storm." *Smithsonian*. August 1979.
Leakey, Richard, and Alan Walker. "Homo Erectus Unearthed." *National Geographic*. November 1985.
Maran, Stephen P. "Sky Reporter." *Natural History*. October 1982.
Marshack, Alexander. "An Ice Age Ancestor." *National Geographic*. October 1988.
Martin, R. D. "Strategies of Reproduction." *Natural History*. November 1975.
McIntyre, Loren. "The Amazon." *National Geographic*. October 1972.

McKean, Kevin, and Mayo Mohs. "Tracking the Killer Waves." *Discover*. August 1983.

Mehringer, Peter J., Jr. "Weapons of Ancient Americans." *National Geographic*. October 1988.

Monastersky, Richard. "Amber Yields Samples of Ancient Air." *Science News*. November 7, 1987.

———. "Bird Fossil Reveals History of Flight." *Science News*. February 13, 1988.

Morell, Virginia. "The Birth of a Heresy." *Discover*. March 1987.

Nicholson, Thomas D. "Total Eclipse." *Natural History*. February 1979.

Norris, Kenneth S. "Tuna Sandwiches Cost at Least 78,000 Porpoise Lives a Year, but There Is Hope." *Smithsonian*. February 1977.

Norsgaard, E. Jaediker. "Connubial Cannibalism." *Natural History*. November 1975.

Oppenheimer, Michael, and Leonie Haimson. "The Comet Syndrome." *Natural History*. December 1980.

Orville, Richard E. "Bolts from the Blue." *Natural History*. June-July 1977.

Overbye, Dennis. "Is Anyone Out There?" *Discover*. March 1982.

Perles, Catherine. "Hearth and Home in the Old Stone Age." *Natural History*. October 1981.

Peterson, Ivars. "Ancient Technology: Pouring a Pyramid." *Science News*. May 26, 1984.

Plazas, Clemencia, and Ana María Falchetti de Sáenz. "Technology of Ancient Colombian Gold." *Natural History*. November 1979.

Poinar, George O., Jr. "Sealed in Amber." *Natural History*. June 1982.

Prospero, Joseph M. "Dust from the Sahara." *Natural History*. May 1979.

Raloff, J. "Captivity Chosen to Save the Ferret." *Science News*. September 8, 1986.

Reeves, Randall R., and Edward Mitchell. "The Whale Behind the Tusk." *Natural History*. August 1981.

Rensberger, Boyce. "Ancestors: A Family Album." *Science Digest*. April 1981.

———. "Bones of Our Ancestors." *Science 84*. April 1984.

Richardson, Louise. "On the Track of the Last Black-Footed Ferrets." *Natural History*. February 1986.

Rigaud, Jean-Philippe. "Treasures from the Ice Age: Lascaux Cave." *National Geographic*. October 1988.

Sax, Joseph L. "America's National Parks: Their Principles, Purposes, and Prospects." *Natural History*. October 1976.

Shipman, Pat. "Baffling Limb on the Family Tree." *Discover*. September 1986.

Sitwell, Nigel. "The 'Queen of Gems'—Always Stunning, and Now More Cultured Than Ever." *Smithsonian*. January 1985.

Sparks, Stephen, and Haraldur Sigurdsson. "The Big Blast at Santorini." *Natural History*. April 1978.

Stewart, John Massey. "A Baby Mammoth, Dead for 40,000 Years, Reveals a Poignant Story." *Smithsonian*. September 1979.

Toon, Owen B., and James B. Pollack. "Volcanoes and the Climate." *Natural History*. January 1977.

Trefil, James. "Stop to Consider the Stones That Fall from the Sky." *Smithsonian*. September 1989.
Trinkaus, Erik. "Hard Times Among the Neanderthals." *Natural History*. December 1978.
Vietmeyer, Noel D. "The Preposterous Puffer." *National Geographic*. August 1984.
Ward, Geoffrey C. "India's Intensifying Dilemma: Can Tigers and People Coexist?" *Smithsonian*. November 1987.
Weaver, Kenneth F. "The Search for Our Ancestors." *National Geographic*. November 1985.
Weinberg, David. "Decline and Fall of the Black-Footed Ferret." *Natural History*. February 1986.
Weisburd, Stefi. "Frost Rings Used to Date Eruptions." *Science News*. January 28, 1984.
Williams, Ted. "The Final Ferret Fiasco." *Audubon*. May 1986.
———. "Long Journey Home of the 'King' of Fish." *Smithsonian*. November 1981.
Zimmerman, David. "The Mosquitoes Are Coming—and That Means Trouble." *Smithsonian*. June 1983.

INDEX